Maths Standard

Elite Internal Assessments

Maths Standard

Elite Internal Assessments

7 Excellent SL IAs for the International Baccaulaureate [IB] Diploma

Zouev-Elite Publishing

This book is printed on acid-free paper.

Copyright © 2018 Zouev Elite Publishing. All rights reserved.

No part of this book may be used or reproduced in any manner whatsoever without written permission, except in the case of brief quotations embodied in critical articles or reviews.

Published 2018

Printed by Lightning Source

ISBN 978-1-9996115-0-7, paperback.

TABLE OF CONTENTS

Introduction	3
IA EXAMPLES	13
1. Modelling Blood Flow in the Central Aorta	15
2. Modelling the Probability of Streaks Within a Season	35
3. Investigating the Koch's Snowflake	49
4. Pigeons, Birthdays, and 607 Students	69
5. Identifying and Comparing Dunkin Donut's	81
6. Ciphers and their Solutions	95
7. The Perfect Dive for a Swimmer	113

Introduction

The mathematics internal assessment provides the student with an opportunity to explore mathematical ideas. The criteria for success assesses the extent to which the student has engaged creatively and critically with mathematics.

Moderators are impressed with students who find an original approach to their subject. This could mean applying mathematics that was developed for one situation to another. An example of this would be applying the mathematics behind the Birthday paradox to Bitcoin addresses. It could also refer to the use of mathematical techniques to investigate everyday objects. For instance, finding the volume and/or surface area of an object using calculus. Creativity can be shown by finding a novel way to represent a mathematical concept. For example, using domino toppling to model the action of logic gates in computers.

The exploration should show development. The research questions posed at the beginning should seek to analyze simple problems and progressively move towards more complex problems as these are solved, getting deeper and deeper within the investigation. It is also important to critically evaluate the findings as one progresses and to assess whether they are accurate or not. An example of this is provided by the following article:

https://education.ti.com/images/online_courses/t3/calculus2/mod28/568-74_nov.pdf
https://goo.gl/STihNt

The student should include reflection on their use of mathematics. This could include justifying any assumptions made by a mathematical model or explaining why a certain statistical test has been used to examine a hypothesis.

The level of mathematics used in the IA is expected to be commensurate with that of the course. Students can use material from their specification but are not restricted to it. The finished exploration should resemble a journal article. Symbols, formulae, equations, identities, relations can be formatted using an equation editor which can be found in Word or Pages. The ambitious may prefer to use LaTeX which is used in the production of technical and scientific documentation.

Presentation	Software
Typesetter	Word, Pages, LaTeX
Data	Excel, Numbers
Graphs	Desmos, Geogebra
Probability distributions	Geogebra
Vector lines and planes	Geogebra

The finished work should be no more than 12 pages (not including the bibliography).

The IA will be internally assessed, and the final mark will constitute 20% of the total grade.

SELECTING A PROJECT

One approach to the IA is the construction of a mathematical model. These attempt to describe and predict real-world phenomena.

The simplest models look for a mathematical relationship between two variables using pre-existing data. For example, showing how the record times of different athletic events (short / long distance running) change over time.

http://mei.org.uk/files/pdf/comprehension/Modelling_Athletic_Records.pdf
https://goo.gl/NGTSn9

Another approach to model building involves collecting your own data.

This may involve setting up a simple experiment. For example, finding a function that describes the relationship between the viscosity of a liquid and temperature.

Experiments may not always be practical or ethical. In such instances the student may prefer to use a simulation as a means of collecting data. For instance, students interested in the mathematics of disease transmission can simulate this by using a pack of playing cards.

http://motivate.maths.org/content/DiseaseDynamics/Activities/26CardEpidemic
https://goo.gl/uJaJX8

Simulations can also be found online. With these apps variables can be manipulated and the results can be recorded. For example, traffic flow could be analysed using a Java app.

http://www.traffic-simulation.de/
https://goo.gl/XhChDh

Data can be analysed by finding a function that describes the relationship between variables. This can be achieved by the method of least squares. This approach uses partial differentiation, a method whose difficulty is commensurate with the specification.

https://www.essie.ufl.edu/~kgurl/Classes/Lect3421/Fall_01/NM5_curve_f01.pdf
https://goo.gl/iZ1Fzo

Software such as Desmos can also be used to analyse data. It is suitable for exploring trigonometric and exponential functions. It is also able to analyse graphs where the function depends upon the domain values.

https://ibmathsresources.com/2017/01/29/maths-of-global-warming-modeling-climate-change/
https://goo.gl/pCEXLG

Mathematical models can also be developed from theory. These have a deeper understanding of phenomena than those models derived from just analysing data.

An excellent explanation of this process has been put together by the American Mathematical Society.

http://www.ams.org/publicoutreach/feature-column/fc-2012-09
https://goo.gl/XCnpEq

A list of mathematical models has been compiled by Plus Magazine:

https://plus.maths.org/issue51/package/MMP_modelling_toolkit.pdf
https://goo.gl/WUxGzY

Some models require the solution of differential equations and so are more suitable for ambitious SL students.
https://www.maa.org/press/periodicals/loci/joma/the-sir-model-for-spread-of-disease-the-differential-equation-model
https://goo.gl/Vfxx6w

Differential equations can be solved using a numerical approach such as Euler's method.

https://www.maa.org/press/periodicals/loci/joma/the-sir-model-for-spread-of-disease-eulers-method-for-systems
https://goo.gl/XnXxcT

These can also be solved by using computer software such as Vensims.
http://vensim.com/building-a-simple-vensim-model/
https://goo.gl/HmxRmx

If models are not your thing then the content specification of your course can suggest other possibilities for exploration.

Topic 1 – Algebra

Investigate the sum to infinity of a geometric sequence using fractals.

https://www.amazon.co.uk/002-Fractals-Classroom-Complex-Mandelbrot/dp/0387977228/ref=pd_sim_14_2?_encoding=UTF8&psc=1&refRID=7Q0VYZ3V68V5F4CSCQHG

Topic 2 – Functions and equations

Investigate equations that have no analytical solution.

http://wwwf.imperial.ac.uk/metric/metric_public/numerical_methods/iteration/fixed_point_iteration.html

Topic 3 – Circular functions and trigonometry

Investigate climate change using a trigonometric model.

https://ibmathsresources.com/2017/01/29/maths-of-global-warming-modeling-climate-change/

Topic 4 -Vectors

Investigate the Knight's journey in chess.

https://nrich.maths.org/1317

Topic 5 – Statistics and Probability

Investigate variations on the birthday paradox. For example, find the number of people required so that the probability of any two of them sharing the same bitcoin address is greater than 50%

https://download.wpsoftware.net/bitcoin-birthday.pdf

Topic 6 – Calculus

Investigate the volume of an everyday object by finding a function(s) that forms the surface of the object when rotated about the x-axis.
There are many interesting areas of maths that can provide material for the IA outside the specification. Here are a few examples:

Random Walks:

These describe pathways composed of random steps in a given number of directions. For instance, random walks along the edges of a polyhedron. This topic may be unfamiliar to the student but is easy to grasp and provides ample opportunities for investigation.

http://www.logic-books.info/sites/default/files/k08-mathematical_circus.pdf

Markov Chains:

These analyse how systems change with time. For example, the movement of share prices on the stock-market.

https://web.wpi.edu/Pubs/E-project/Available/E-project-031411-153131/unrestricted/MQP_HS1_3699.pdf

They can also be used to show how card tricks work, for example the Kruskal Count:

http://www.osaka-ue.ac.jp/zemi/nishiyama/math2010/kruskal.pdf

Chaos:

Recursive formulae can be used to explore so-called chaotic phenomena. These refer to systems whose behaviours can vary radically depending on small changes in starting conditions. One such recursive formulae - the Verhulst equation - looks at how fluctuations in populations alter when a fixed limit to growth is introduced.

http://mei.org.uk/files/papers/All%20packs/MEI_C4b.pdf (pages 10-12)

Assessment Criteria:

Criterion A: Communication (4 marks)

This criterion assesses the student's ability to communicate their understanding of the mathematical ideas they have chosen to explore in the appropriate form. To achieve full-marks the IA must meet the following requirements:
Well-organized: An IA should be structured like a journal article:

Introduction to the topic.
Aim / approach to the problem / why topic was chosen / real-life applications.
Formulation and exploration of research question (s).
Conclusion summary of the results, limitations of the assessment, possible future avenues of enquiry.
Bibliography: this must list all sources (books, journals, websites, personal communications) you consulted when writing your IA. Occurrences of these sources in the text must be referenced and linked to their relevant location in the bibliography.
Appendix: this consists of data, critical values used in hypothesis testing, additional material that did not fit into the into main text.

Coherent: Displays a logical development and can be understood by fellow students. Make sure your explanations are clear and include diagrams if they help the reader to follow your working.

Concise: Material irrelevant to the aim set out in the introduction is to be avoided. Analysis that are not immediately relevant to the aim of the IA can be included in the appendix. Try not to repeat yourself, if you show a particular calculation that you do several times over the investigation, do not write it again every time but refer to the example shown and mention that the same method was followed.

Complete: The aim of the IA has been met. All concepts, ideas, formulae, etc. have been clearly described. Keeping a balance between a complete and a concise IA might be difficult, so make sure you identify the key concepts in your IA and that you explain them well, avoiding explanations which are not relevant.

The introduction should clearly state what the aim of the mathematical investigation is and why this topic was chosen (there is also room here to show personal relevance!). Moreover, the conclusion should be concise and summarize the relevant findings of the exploration, answering the initial aim of the investigation.

Criterion B: Mathematical Presentation (3 marks)

This criterion assesses the student's ability to use different forms of mathematical representation such as formulae, equations, diagrams, tables, and graphs. These should only be included when relevant for the investigation and the student should refer to them in the main text. Any additional graphs or tables that are not essential to follow the main discussion should be included in the appendix. Graphs and tables must be labeled, including the relevant units when appropriate.

Mathematical language (notation, symbols and terminology) should be used when communicating mathematical ideas, reasoning and findings in a correct way. All the variables and key terms used should be defined in a simple and concise way.

Appropriate technology such as graphic display calculators; and software such as equation editors, spreadsheets, dynamic geometry, computer algebra, and typesetting software such as LaTeX could be used to enhance the presentation of mathematics. Finally, a complete bibliography should be included.

Criterion C: Personal Engagement (4 marks)

This criterion assesses the student's inventiveness and investment in the exploration. It could involve finding novel ways to approach a topic, exploring unfamiliar mathematics or comparing different methods. For example, setting up a simulation to model a phenomenon. It could also refer to how reflection on the topic has stimulated questions and conjectures which are examined in the text.

Personal engagement can be shown in the introduction by explaining why the student chose a particular topic or by using real data and applying mathematics to real-life situations. Moreover, the way the task is approached by using examples and approaches created by the student will also demonstrate a good level of personal engagement.

Criterion D: Reflection (3 marks)

This criterion assesses the student's ability to provide a critical commentary throughout their IA. Reflection seeks to provide justification for mathematical decisions made in the exploration. For example, students developing mathematical models should state the assumptions they have made and provide a rationale for using them. Projects that involve statistics should explain the choice of calculation or tests used to explore data.

The student can show reflection throughout the text by critically analyzing the results obtained as the exploration progresses. This could mean that the student reflects on the accuracy of his/her results and explains the differences between the mathematical model used and the real-life results. It is very important that the validity of the models are evaluated thoughtfully. In the conclusion reflection may be demonstrated by considering limitations or by suggesting avenues for further research.

Criterion E: Use of Mathematics (6 marks)

This criterion assesses the student's ability to apply mathematics that is commensurate with the level of their course specification. Mathematics taken from outside the specification is expected to be of a similar level. The mathematical arguments should be logical and avoid gaps in reasoning. Numerical computations are required to be accurate and given to an appropriate level of accuracy. The techniques employed in the exploration will be assessed to ensure that they are used correctly. Instances of this include the appropriate differentiation or integration of a function or the selection of the relevant statistical hypothesis test. It is very important that the student demonstrates a good understanding of the mathematics used by applying then correctly and explaining the key concepts.

Teacher Feedback:

The student will be required to submit a complete draft of their IA – containing an introduction, conclusion and all planned content to sufficiently address all five criteria. They will receive feedback on their draft and then be given an opportunity to revise it to submit a final version.

'I confirm that this work is my own and is the final version. I have acknowledged each use of the words or ideas of another person, whether written, oral or visual.'

Resources:

Software:

Wolfram Alpha:
Limits. Calculus. Equations. Matrices.
https://www.wolframalpha.com/examples/math/

Desmos:
Graphing calculator.
https://www.desmos.com/

Geogebra:
Coordinate geometry in 2D and 3D. Probability distributions. Calculus.
https://www.geogebra.org/

Vensim:
Mathematical modelling using differential equations.
http://vensim.com/

R:
Using R to visualise data:
http://flowingdata.com/category/tutorials/

Graphic Display Calculators:

These can be linked to PCs which allow for output to be copied to documents.
Texas Instruments:
https://education.ti.com/en/software/search
CASIO:
https://edu.casio.com/dl/
GDCs are also programmable. For example, they can be programmed to show the result of iterative calculations.

Web sites:

Ideas:
Ideas with hyperlinks.
https://ibmathsresources.com/

Maths Investigation Ideas for A-level, IB and Gifted GCSE Students.
http://www.cobblearning.net/campbellmathkyleturner/files/2016/08/Maths_Investigation_Ideas-2msw9dj.pdf

200+ topics.
https://www.fairviewhs.org/staff/emily-silverman/classes/ib-math-hl/files/30763

MEI structured exam questions contain ideas for projects. Refer to the comprehension section accompanying each exam paper.

http://mei.org.uk/files/papers/All%20packs/MEI_C4b.pdf

Geometry Junkyard:
https://www.ics.uci.edu/~eppstein/junkyard/

Wolfram Mathworld:
'The web's most extensive mathematics resource'
http://mathworld.wolfram.com/

Martin Gardener:
These books contain many ideas suitable for projects of all levels. Some of his books can be found as pdfs on the internet.
Highly recommended are Mathematical Circus and Hexaflexagons and other mathematical diversions:
http://www.logic-books.info/sites/default/files/k08-mathematical_circus.pdf
http://geofhagopian.net/papers/GARDNER01.pdf

Dr Maths:
http://mathforum.org/library/drmath/view/56711.html

Maths Explorers Club:
http://www.math.cornell.edu/~mec/

Plus Magazine:
https://plus.maths.org/content/Package

Popular topics
Epidemiology:
https://www.maa.org/press/periodicals/loci/joma/the-sir-model-for-spread-of-disease-the-differential-equation-model
http://motivate.maths.org/content/DiseaseDynamics/
https://plus.maths.org/content/mathematics-diseases

Cosmology:
Resources from NASA:
https://spacemath.gsfc.nasa.gov/books.html
Exoplanet orbit database:
http://exoplanets.org/

Games theory:
Strategies and mathematics of 2-player games
www.ics.uci.edu/~eppstein/cgt/

You Tube Channels

Numberphile:
https://www.youtube.com/user/numberphile
Vihart:
https://www.youtube.com/channel/UCOGeU-1Fig3rrDjhm9Zs_wg

7 EXAMPLES OF

EXCELLENT INTERNAL ASSESSMENTS

The commentaries featured in this section are all recently submitted IAs that scored very highly after being moderated by the IBO. To prevent plagiarism and duplication of results, the appendices have been omitted. We do not retain the copyright of these commentaries, nor is this publication endorsed by the IBO. The Internal Assessments are being re-printed with the permission of the original authors.

1. Modelling Blood Flow in the Central Aorta

(assessment extracted from PDF)

Modelling Blood Flow in the Central Aorta

Introduction

I harbour a deep love for biology and thus, when thinking about topics for my math IA, ideas from biology came to me naturally. I narrowed down on one aim: to model the blood flow in the central aorta, the body's main artery. In order to achieve this aim, I will:

- derive Poiseuille's law for laminar flow,
- utilise Poiseuille's formula to calculate blood flow values for a data set of aortic pressure values measured during a heart cycle,
- find an appropriate mathematical equation that models the calculated blood flow values, and
- test the closeness of fit and the accuracy of the model for another heart cycle from the same data set.

The significance of the blood flow in the human body is truly fascinating in its organisation and specific mechanisms to create the framework for life. This was very interesting for me as I considered how efficiently blood flow took place in our bodies. From this thought, I wondered if the cyclic, continuous blood flow could be modelled against time. I believe that such a model could lend itself to important applications, particularly in personalised medicine where the creation of detection softwares would improve monitoring and predictions for cardiovascular disease. For example, it is possible that such a model could even be adapted to the coronary arteries and inform heart patients faster if they are about to have a heart attack! This interesting topic and many useful applications encouraged me to begin this investigation. I am further prompted towards pursuing it because of my personal experiences of having family members suffering from heart disease and the possibility that advancements in this area of study have the potential to make life easier for heart patients.

Poiseuille's Law

J. L. Poiseuille (1799–1869) was a French scientist who investigated the movement of fluids through cylindrical structures. He postulated that fluids move in laminae or 'sheets' that are stacked on one another which all moved at different velocities. Two representations of the movement of a fluid by his law are given below. *Figure 1* is a three-dimensional diagram showcasing the velocity of the fluid at varying distances from the tube wall. It is seen that the velocity follows a continuous variation and is greatest at the centre of the tube, gradually decreasing to zero at the tube walls. The laminae in this representation are very thin and thus not decipherable from the figure. *Figure 2* showcases the laminae more conspicuously, representing once more, the variation in velocity of the laminae. The reason for this variation of velocity becomes apparent when this figure is viewed as a lever: less force is required to move a lever with a greater radius and for the same force exerted uniformly across the lumen of a tube, fluid at a greater distance from the tube wall travels with a greater velocity.

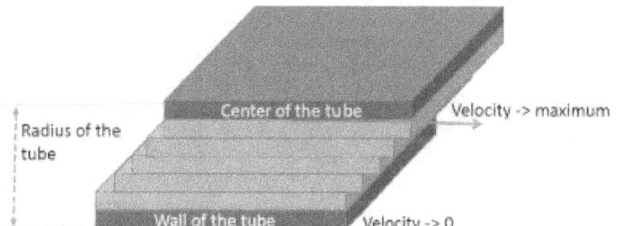

Figure 1 A three-dimensional representation of fluid movement through a tube, as made by me on GeoGebra. The junction of the striped and black area represents a section of the tube wall. The protruding volume represents flowing blood.

Figure 2 A representation of fluid movement through half a tube, as made by me on PowerPoint 15.13.1. The bottom represents the tube wall and the top represents the centre of lumen. Laminae are clearly visible.

Poiseuille's Law for laminar flow formulates the flow rate of a fluid through a tube. The derivations for this law are given below. Some of explanations for the derivation have been omitted due to their basis in physics and not mathematics.

$Pressure = \frac{Force}{Area}$

$Force = Pressure \times Area$ ...(1)

Equation (1) is a standard equation in physics for the relationship between force, pressure and area. Since force is required to propel a fluid through a tube and a pressure difference allows its movement from one location to another, this formula is befitting to this derivation. Force is shortened to 'F', pressure to 'P' and area to 'A' for use henceforth. While considering the forces that cause a fluid to move through a tube, it is important to consider two major force types: forces that propel the tube forward or the 'driving' force and the forces that resist its flow or the 'resistive' forces.

$$F_{driving} = (P_1 - P_2)\pi r^2 \quad ...(2)$$

$$F_{resistive} = -\eta A \frac{dv}{dr} = -\eta(2\pi rL)\frac{dv}{dr} \quad ...(3)$$

Key terms:

$F_{resistive}$			Resistive force
$F_{driving}$	Driving force	η	Viscosity
P_1	High pressure	A	Surface area of the fluid
P_2	Low pressure	$\frac{dv}{dr}$	Rate of change of velocity with radius
r	Radius of the tube	L	Length of the tube

Equation (2) and (3) are derived from equation (1). In equation (2), the pressure difference between two given points in a tube causes the fluid to move from the area of higher pressure to the area of lower pressure and hence, P is replaced by the difference in pressures across the two points. The cross-sectional area or the area of a circle of no width is given by πr^2 and replaces A. In equation (3), the negative sign represents that the resistive force acts in the opposite direction of the driving force. Viscosity of a fluid, represented by η, is a property of fluids that causes resistance between moving laminae. The viscosity is an adjusting factor of the resistive force and thus important to equation (3). A is retained from equation (1), and in this case represents the surface area of the fluid moving through the tube, since friction acts on the exposed surfaces of a given cylinder of fluid. Since the movement of the fluid is continuous and does not end at any point on the considered portion of the tube, its surface area is considered to be that of a hollow cylinder, or $2\pi rL$. In laminar flow, the velocity of the fluid at different points in the tube is different. Thus, the rate of change of velocity or v with respect to radius r, or $\frac{dv}{dr}$ is also considered in this formula. At a steady state or equilibrium, the driving and resistive forces are equal to each other, as shown below.

$$F_{driving} = F_{resistive}$$

$$(P_1 - P_2)\pi r^2 = -\eta(2\pi rL)\frac{dv}{dr}$$

This is a first order differential equation and its rearrangement to make $\frac{dv}{dr}$ the subject allows us to then bring it into the variable separable form which we studied in the Calculus Option. The subsequent integration of this equation allows us to find -v as in Equation (5). In order to solve for c, the theory discussed previously regarding the velocity at the tube walls is used and a complete equation for velocity through a tube for a fluid undergoing laminar flow is found as seen in Equation (6).

$$-\frac{dv}{dr} = \frac{(P_1-P_2)r}{2\eta L} \dots (4)$$

$$-\int \frac{dv}{dr} = \int \frac{(P_1-P_2)r}{2\eta L}$$

$$-\int dv = \frac{(P_1-P_2)}{2\eta L} \int r\, dr$$

$$-v = \frac{(P_1-P_2)}{2\eta L} \times \frac{r^2}{2} + c \dots (5)$$

Finding c using $r = R$ when $v = 0$:

$$0 = \frac{(P_1-P_2)R^2}{4\eta L} + c$$

$$c = -\frac{(P_1-P_2)R^2}{4\eta L}$$

$$v = \frac{(P_1-P_2)}{4\eta L} \times (R^2 - r^2) \dots (6)$$

Poiseuille's Law is particularly useful in the field of medicine as it is used to determine the volume of fluid transported or delivered, including in blood flow in the arteries and during intravenous administration into the veins. In each of these applications, one of the important applications of this derivation is finding the volume of fluid flowing through the tube with time. To do this, a standard derivation exists, but I felt like a Riemann Sums, which is a topic we covered in our Calculus Option would be well-suited for this purpose. The approach is to find the change of volume with time of blood using the velocity function derived above and the cross sectional area of the fluid using Riemann Sums.

Riemann Sums are a method of dividing a closed interval into several small sub-intervals, adding the sub-intervals together and taking the limit where the width of the subintervals tends to zero. This produces an integral which has many applications in modelling, particularly for continuous variables. Reimann upper and lower sums can be calculated depending on which value within the sub-interval is used in the calculation. The difference between the upper and lower sums is negligible as the width of the sub-interval decreases. A graphical representation of Riemann Sums is shown in *Figure 3*.

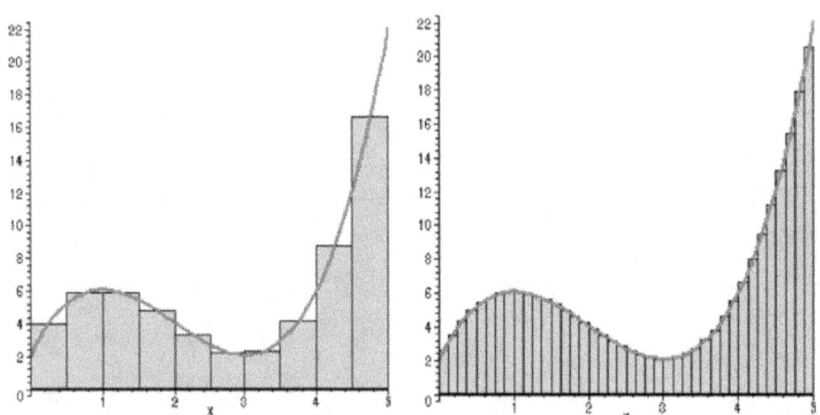

Figure 3 A graphical representation of producing an integral using Riemann Sums method.
[Adapted from: "Riemann Sums and Numerical Integration." *SDSU*]

Both the graphs in Figure 3 show a curve (the red line) on the interval $0 < x < 5$. The width of the sub-interval is different on each graph. The first graph features sub-intervals of width 0.5 units and the second graph 0.125 units. On each graph, the integral was estimated by adding together the sum of the rectangles made by the sub-intervals. However, it can be noticed that in the second graph, the usage of smaller sub-intervals allowed us to make a closer estimation of the integral. Thus, for accuracy, the sub-intervals are passed to the limit such that the sub-interval width tends to zero.

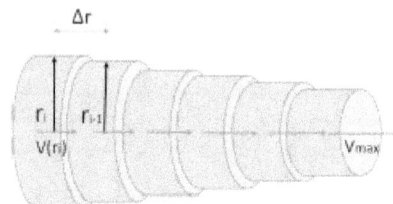

Figure 4 A three-dimensional representation of the visualisation used to find the volume of fluid per unit time using Riemann's Sum method.

To employ the Riemann Sums method for my purpose, the visualisation in *Figure 4* is utilised. The volume of each ring is computed using the formula of the volume of a cylinder = $2\pi r h$ where r is the radius and h is the height. This standard equation is

$$\text{area of ring} = 2\pi r_i \Delta r \quad \ldots(7)$$

where $\Delta r = r_i - r_{i-1}$

modified for this purpose as shown in equation (7).

The product of the volume of the ring and velocity is found and a sigma function is employed to denote the sum of this product from the first to the nth sub-interval. To obtain accuracy in the result, the calculation for the number of sub-intervals n tending to infinity was used in the limit in Equation (9). The approach of finding the product of velocity and the area of the ring is justified given the following reasoning. As Δr tends to zero, the volume of the ring tends towards the cross sectional area of the ring. Hence, by a dimensional analysis, multiplying velocity, which is length over time, and the cross-sectional area gives us volume over time. The sigma notation is used to represent the sum of individual rings as shown in Equations (8) and (9).

$$2\pi r_i \Delta r v(r_i) = 2\pi r_i v(r_i) \Delta r$$

$$\sum_{i=1}^{n} 2\pi r_i v(r_i) \Delta r \quad \ldots(8)$$

$$flux = \lim_{n \to \infty} \sum_{i=1}^{n} 2\pi r_i v(r_i) \Delta r \quad \ldots(9)$$

An infinite number of sub-intervals necessitates a very small width of each sub-interval or Δr which increases the accuracy of the integral estimation as discussed previously using *Figure 3*. This Riemann Sum tends itself to the integral shown below as we try to find the sum of infinitesimally small sub-intervals within the closed interval 0 to R. I use the integral and omit the Reimann Sum calculations since the former is much more accurate. Integration allows us to find $\frac{dV}{dt}$ or the **rate of change of volume with time: the flux**. The upper and lower bounds of the integration are considered to be zero and the maximum radius R, which are the limits of the closed interval considered. The final equation is presented in Equation (10).

$$\int_0^R 2\pi r v(r) dr$$

$$\int_0^R 2\pi r \frac{(P_1-P_2)}{4\eta L} \times (R^2 - r^2) dr$$

$$\frac{\pi(P_1-P_2)}{2\eta L} \times \left[\frac{R^2 r^2}{2} - \frac{r^4}{4}\right]_{r=0}^{r=R}$$

$$Flux = \frac{\pi(P_1-P_2)R^4}{8\eta L} \quad ...(10)$$

Note that in these calculations, $v(r_i)$ represents velocity as a function of radius. Equation (10) equation is representative of the change in volume with time ($\frac{dV}{dt}$) or flux.

It is interesting to note the relationship between the flux and radius of a fluid moving through a tube. The fourth power relationship of radius to flux implies that small changes in the radius of the tube result in exponential changes in the flux of a fluid. This fact is very pertinent to the many applications that the results of this investigation tend to. For example, in applying Poiseuille's law to the flux of blood in the body's main artery, the aorta, it is noticed that aortic stenosis—a process that causes the narrowing of the aorta—deters blood flux significantly and results in a strain to the heart chamber pumping it out as the heart tries to accommodate for this narrowing. Severe aortic stenosis often implicates high mortality for the patient.

Flux Calculation and Modelling for the Central Aorta

In calculating the flux of blood moving through the central aorta, the viscosity of blood and the length of the aorta considered were constant parameters. A value of **0.00001563 mmHg**.s was used for the **viscosity** from the data table provided by OpenStax (n.p.). The radius of the aorta varies slightly with pressure and distance from the heart. Hence, for the sake of simplicity in this investigation, it was assumed that the radius was constant at all pressures. For the length, only the distance of the central aorta was considered, a range within which there was no significant change in the radius due to narrowing of the vessel (Erbel 137). The **length of the aorta** was taken to be **0.208 m** and its **radius** to be **0.01056 m** in a normal, healthy human (Dotter 916). This is summarised in the table below.

Table of constant parameter values

Viscosity	0.00001563 mmHg.s
Length	0.208 m
Radius	0.01056 m

A data table from a haemodynamics study by Zhang et al was utilised to find pressure values in the aorta against time. *Table 1* on the left and *Graph 1* below represent the data collected and plotted respectively. Consequently, the flux was calculated using which $Flux = \frac{dV}{dt} = \frac{\pi(P_1 - P_2)R^4}{8\eta L}$ is Equation (10). Values from this data set and the aforementioned parameters were used.

Table 1. Time and Aortic Pressure from data set

Time/s	Aortic Pressure/mm Hg
4.00	84.00
4.17	71.00
4.20	80.00
4.24	105.00
4.30	109.00
4.40	117.00
4.42	114.00
4.46	101.00
4.60	91.00
4.80	72.00
4.86	100.00
4.90	104.00
5.00	114.00
5.07	116.00
5.18	100.00
5.20	92.00
5.40	75.00
5.44	73.00
5.58	104.00
5.60	112.00
5.70	118.00
5.75	121.00
5.82	104.00
5.80	100.00
6.00	85.00

Graph 1. Aortic Pressure with time

For the flux calculations, the values of the constant parameters and the time values in the data set were maintained in Standard International units. However, the pressure was given in mmHg as in the data set since this is the more common unit for measuring pressure in the human body and this was maintained through the calculations as well. Equation (10) discussed above was used for this calculation and the obtained flux values were graphed against time as shown in *Table 2* and *Graph 2*. Note that since the instantaneous flux, or the flux at a given point in time, is being found, the fluid movement is considered to take place over a distance *d* where $d \to 0$. Under this condition,

P_1-$P_2 \approx P_1$ in Equation (10). This approximation is used in the flux calculations. For all calculations, the spreadsheet software Microsoft Excel version 15.13.1 was utilised.

Table 2. Time and Calculated Aortic Flux

Time/s	Flux
4.00	0.005997
4.17	0.006252
4.20	0.006297
4.24	0.006357
4.30	0.006447
4.40	0.006597
4.42	0.006627
4.46	0.006687
4.60	0.006897
4.80	0.007196
4.86	0.007286
4.90	0.007346
5.00	0.007496
5.07	0.007601
5.18	0.007766
5.20	0.007796
5.40	0.008096
5.44	0.008156
5.58	0.008366
5.60	0.008396
5.70	0.008546
5.75	0.008621
5.82	0.008726
5.80	0.008696
6.00	0.008995

Graph 2. Aortic Flux with time

In *Graph 2*, the flux appears to follow a cyclic wave pattern, drawing parallels to the trigonometric sine and cosine waves. As the heat beats at a steady pace, the flux changes with it, and there exists a constant period for the change in flux with time. Considering its wave-like fluctuation and periodicity, I wondered if this curve could be represented as a sinusoidal function. When we studied trigonometry in class, we graphed compound trigonometric functions consisting of sine and cosine waves. Then, I noticed that the resulting wave was a hybrid of its constituent waves. This learning made me wonder whether I could mathematically fit the flux curve to a trigonometric function that consisted of sine and cosine functions to create a close fit to the unique shape of this graph. While researching this, I came across the Fourier series, a mathematical tool for the representation of periodic functions as a sum of infinite sine and cosine functions. Presented below is an explained derivation of the Fourier series formula that I propose, ending with the general formula of a Fourier series. We start off with the general formulae for the sine and cosine wave functions as shown below.

$y_1 = a\,sin(\frac{2\pi}{T}x) + c$

$y_2 = b\,cos(\frac{2\pi}{T}x) + d$

where $\frac{2\pi}{T} = \omega$

In the formulae above, a and b are the amplitudes of the sine and cosine waves respectively. $\frac{2\pi}{T}$ or ω is the angular frequency, a factor maintaining the period of wave since the time period T is generally denoted as $\frac{2\pi}{n}$ where n is a factor inversely proportional to the period. Lastly, c and d are the vertical translations. In order to create a compound trigonometric function of either sine or cosine waves, many different sine and cosine wave functions would have to be added up together to create a compound trigonometric function. For this, a series of sine and cosine waves are chosen as seen below.

$$a_1 sin(\tfrac{2\pi}{T}x) + a_2 sin(2\tfrac{2\pi}{T}x) + a_3 sin(3\tfrac{2\pi}{T}x) + \ldots a_n sin(n\tfrac{2\pi}{T}x) + c$$

$$b_1 cos(\tfrac{2\pi}{T}x) + b_2 cos(2\tfrac{2\pi}{T}x) + b_3 cos(3\tfrac{2\pi}{T}x) + \ldots b_n cos(n\tfrac{2\pi}{T}x) + d$$

where n = 1, 2, 3...

Each term in the series has a progressively smaller period due to the increasing numerator by a factor of *n* (caused by multiplying ω into *n*) and modify the overall shape of the compound curve. The variation of their individual amplitudes (a_n and b_n) also play a role in determining the shape of the modelled wave. Together, these two factors allow a closely fit model of the blood flow in the central aorta to be created. Using the sigma notation, these series can be represented comprehensively:

$$\sum_{a=1}^{\infty} (a_n sin(n\tfrac{2\pi}{T}x)) + c$$

$$\sum_{b=1}^{\infty} (b_n cos(n\tfrac{2\pi}{T}x)) + d$$

Generally for Fourier series based modelling, the function is first classified as odd or even after which the sine or cosine functions given above are picked for use in modelling. However, my function was not an ideal function such as a triangular or sawtooth wave function which are more easily classifiable as odd or even. Thus, I didn't want to assume the nature of my model. Moreover, the presence of more coefficients would improve the accuracy of my model in the situation where the coefficients do not have values of zero (in which case their corresponding terms get cancelled out). I chose to use the Fourier series equation below.

$$y = a_0 + \sum_{n=1}^{\infty} (a_n sin(n\tfrac{2\pi}{T}x) + b_n cos(n\tfrac{2\pi}{T}x))$$

In the formula a_0 represents the vertical translation of the function, a_n is the coefficient of the sine functions and b_n is the coefficient of the cosine functions. These factors adjust the amplitudes of the wave (since this compound function will have multiple amplitudes). T accounts for the average period of the function, which in the case of my data set is 0.63 seconds and so $\frac{2\pi}{T}$ is constant and has a value of $9.977s^{-1}$. This constant adjusts the period to that of the flux wave. The sigmoid function ranges from n=1 to infinity and represents an infinite sum of sine and cosine functions. For the purposes of my investigation, I considered n from 1 to 5, where n reaches a high enough value to obtain a closely fit model. This was done since I cannot compute values of n until infinity, only decide a point that represents or approximates infinity. This assumption gave me the equation given below:

$$y = a_0 + a_1 sin(\tfrac{2\pi}{T}x) + a_2 sin(2\times\tfrac{2\pi}{T}x) + a_3 sin(3\times\tfrac{2\pi}{T}x) + a_4 sin(4\times\tfrac{2\pi}{T}x) + a_5 sin(5\times\tfrac{2\pi}{T}x)$$
$$+ b_1 sin(\tfrac{2\pi}{T}x) + b_2 sin(2\times\tfrac{2\pi}{T}x) + b_3 sin(3\times\tfrac{2\pi}{T}x) + b_4 sin(4\times\tfrac{2\pi}{T}x) + b_5 sin(5\times\tfrac{2\pi}{T}x) \quad \ldots(11)$$

The independent variable or x in my investigation was time and the dependent variable or y was the flux. This equation contains 11 constant parameters, namely a_0, a_1, a_2, a_3, a_4, a_5, b_1, b_2, b_3, b_4 and b_5. In order to find the values of the 11 constant parameters, I created 11 simultaneous equations and decided to solve them by the matrix method.

Since matrices are not in the IB Mathematics HL syllabus, I'll first discuss some relevant theory about them. A matrix, in general terms, is a rectangular array of numbers that is arranged in rows and columns (University of Surrey). Oftentimes, they mathematically represent a system of linear equations and also allow for multiple simultaneous equations to be solved with relative ease. An important concept in matrices is that of inverse matrices. Consider a matrix Mat A and its inverse matrix Mat A^{-1}. Mat A^{-1} is a unique matrix that when multiplied into Mat A gives the identity matrix. The identity matrix is analogous to the number 1 in linear equations: it does not change the magnitude of the matrix that it is multiplied into. Every matrix which has an equal number of rows and columns (called a square matrix, where none of the rows or columns are fully zero) and which has a determinant that is not equal to zero has an inverse matrix. The method to find the inverse matrix varies depending on the number of rows and columns in the matrix. An example of the method to find the inverse matrix for a 2×2 matrix is given. First the determinant of the matrix is found. For a 2×2 matrix, the determinant = $\frac{1}{ad-bc}$

when Matrix A = $\begin{bmatrix} a & b \\ c & d \end{bmatrix}$

Then the b and c terms are multiplied into -1 and the a and d terms are transposed. The determinant is then multiplied into this modified matrix. The result is the inverse matrix.

modify Matrix A to $\begin{bmatrix} d & -b \\ -c & a \end{bmatrix}$

inverse of Matrix A = $\frac{1}{ad-bc} \times \begin{bmatrix} d & -b \\ -c & a \end{bmatrix}$

This ingenious method has some very useful applications to the solving simultaneous equations. Consider an equation is in the form shown below where Mat A and Mat Y contain known values and Mat X is the matrix whose values need to be found. Multiplying Mat A^{-1} into the equation eliminates Mat A from the LHS, leaving only Mat X (multiplication of Mat X into the identity matrix simply gives us Mat X). Therefore, the values contained in Mat X are found on the RHS by the multiplication of Mat A^{-1} and Mat Y.

Mat A × *Mat X* = *Mat Y*

Mat A^{-1} × *Mat A* × *Mat X* = *Mat A^{-1}* × *Mat Y*

Mat X = *Mat A^{-1}* × *Mat Y*

This theory in mind, I created a matrix setup in the same format as outlined. Values from equation (11) and 11 data points from my Time-Flux data set were used for this. Time, or x, values were used in Matrix A and flux, or y, values were used in Matrix Y. The coefficients constituted Matrix X. These matrices are given in the next page and will henceforth be referred to as Matrix A, Matrix X and Matrix Y respectively.

$$\begin{bmatrix} 1 & \sin(\tfrac{2\pi}{T}x_1) & \sin(\tfrac{4\pi}{T}x_1) & \ldots & \sin(\tfrac{10\pi}{T}x_1) & \cos(\tfrac{2\pi}{T}x_1) & \cos(\tfrac{4\pi}{T}x_1) & \ldots & \cos(\tfrac{10\pi}{T}x_1) \\ 1 & \sin(\tfrac{2\pi}{T}x_2) & \sin(\tfrac{4\pi}{T}x_2) & \ldots & \sin(\tfrac{10\pi}{T}x_2) & \cos(\tfrac{2\pi}{T}x_2) & \cos(\tfrac{4\pi}{T}x_2) & \ldots & \cos(\tfrac{10\pi}{T}x_2) \\ 1 & \sin(\tfrac{2\pi}{T}x_3) & \sin(\tfrac{4\pi}{T}x_3) & \ldots & \sin(\tfrac{10\pi}{T}x_3) & \cos(\tfrac{2\pi}{T}x_3) & \cos(\tfrac{4\pi}{T}x_3) & \ldots & \cos(\tfrac{10\pi}{T}x_3) \\ \vdots & & & & & & & & \\ 1 & \sin(\tfrac{2\pi}{T}x_9) & \sin(\tfrac{4\pi}{T}x_9) & \ldots & \sin(\tfrac{10\pi}{T}x_9) & \cos(\tfrac{2\pi}{T}x_9) & \cos(\tfrac{4\pi}{T}x_9) & \ldots & \cos(\tfrac{10\pi}{T}x_9) \\ 1 & \sin(\tfrac{2\pi}{T}x_{10}) & \sin(\tfrac{4\pi}{T}x_{10}) & \ldots & \sin(\tfrac{10\pi}{T}x_{10}) & \cos(\tfrac{2\pi}{T}x_{10}) & \cos(\tfrac{4\pi}{T}x_{10}) & \ldots & \cos(\tfrac{10\pi}{T}x_{10}) \\ 1 & \sin(\tfrac{2\pi}{T}x_{11}) & \sin(\tfrac{4\pi}{T}x_{11}) & \ldots & \sin(\tfrac{10\pi}{T}x_{11}) & \cos(\tfrac{2\pi}{T}x_{11}) & \cos(\tfrac{4\pi}{T}x_{11}) & \ldots & \cos(\tfrac{10\pi}{T}x_{11}) \end{bmatrix} \times \begin{bmatrix} a_0 \\ a_1 \\ a_2 \\ a_3 \\ a_4 \\ a_5 \\ b_1 \\ b_2 \\ b_3 \\ b_4 \\ b_5 \end{bmatrix} = \begin{bmatrix} y_1 \\ y_2 \\ y_3 \\ y_4 \\ y_5 \\ y_6 \\ y_7 \\ y_8 \\ y_9 \\ y_{10} \\ y_{11} \end{bmatrix}$$

Matrix A had to be an 11×11 matrix to allow its multiplication into Matrix X and ensure its inverse could be found. Since a_0 is independent of the time (it is not multiplied by time, or x, in the formula), a column of 1's was made in Matrix A to ensure the correct creation of the matrix. The multiplication of Matrix A and Matrix X gives Matrix Y. Since Matrix X is to be found, the inverse of Matrix A is found and multiplied into Matrix Y as shown below.

The inverse of Matrix A was first computed using 11 consecutive time values from the data set. These values were computed according to the format outlined previously to form all the elements of Matrix A. The spreadsheet software Microsoft Excel version 15.13.1 was used for calculations. Then, the inverse of Matrix A was calculated using the online software BlueBit by .NET Matrix Library 6.1. The GDC too could've been used for this but since all my calculations were on a spreadsheet and the matrix results would be needed again for graphing on the spreadsheet, I decided that it was more efficient to have all my calculations on my laptop. I was unfortunately unable to use the matrix notation with square brackets in this software.

```
Matrix A
1.00000   0.80246  -0.95766   0.34043   0.55140  -0.99847  -0.59671  -0.28789   0.94027  -0.83424   0.05532
1.00000  -0.69235   0.99915  -0.74955   0.08255   0.63042  -0.72156   0.04131   0.66195  -0.99659   0.77625
1.00000  -0.87433   0.84868   0.05056  -0.89775   0.82085  -0.48533  -0.52891   0.99872  -0.44050  -0.57114
1.00000  -0.99421   0.21363   0.94831  -0.41740  -0.85862  -0.10744  -0.97691   0.31735   0.90872  -0.51262
1.00000  -0.88186  -0.83160   0.09766   0.92370   0.77339   0.47150  -0.55537  -0.99522  -0.38313   0.63393
1.00000  -0.08198  -0.16340  -0.24373  -0.32242  -0.39893   0.99663   0.98656   0.96984   0.94660   0.91698
1.00000   0.11721   0.23280   0.34518   0.45280   0.55418   0.99311   0.97253   0.93854   0.89161   0.83240
1.00000   0.49390   0.85891   0.99978   0.87974   0.53012   0.86952   0.51213   0.02109  -0.47545  -0.84792
1.00000   0.94189  -0.63282  -0.51671   0.97999  -0.14172  -0.33594  -0.77429   0.85616   0.19907  -0.98991
1.00000  -0.69417   0.99934  -0.74450   0.07246   0.64019  -0.71981   0.03626   0.66762  -0.99737   0.76822
1.00000  -0.97908   0.39841   0.81696  -0.73085  -0.51957  -0.20346  -0.91721   0.57669   0.68254  -0.85443

Inverse of Matrix A
 0.07174   737.85575    22.37069    22.67552   -0.81077    1.23635   -1.30942    0.54016    0.14245  -748.93964   -32.83283
-0.04710   750.73987    25.76373    29.06691   -1.57549    1.41378   -1.76298    0.76043    0.37959  -763.28805   -41.45069
-0.38033    40.75396     2.74188     5.24268   -0.59513    0.31852   -0.56195    0.46915    0.16348   -41.53144    -6.62081
 0.37130   109.25111     3.37734     3.39005   -0.16151   -0.42261    0.41997    0.03677   -0.38421  -111.16321    -4.71499
-0.01223  -602.44703   -20.26502   -21.42918    1.01418   -1.38601    1.57087   -0.41636    0.05514   612.44459    30.87105
-0.05386   261.46243     8.36914     7.75109   -0.19251   -0.07768    0.13170   -0.05597    0.04762  -265.69189   -11.69007
-0.36310  -717.16665   -18.85598   -16.49910    0.70524    0.18611   -0.04187    0.24964    0.18232   726.89597    24.70743
 0.33730   235.95481     2.72675    -1.84052    0.22218   -0.32897    0.71321   -0.23265   -0.47899  -238.02557     0.95245
 0.14367  -822.13215   -24.72567   -25.44121    0.92179   -1.15976    1.59343   -0.58831   -0.01229   834.56317    36.83733
-0.20731   409.66173    11.59606    11.52096   -0.54780    0.18568   -0.02125   -0.10669    0.19055  -415.68684   -16.58508
 0.01454   172.92196     7.54993    10.47013   -0.54779    0.47855   -0.49370    0.11290   -0.02082  -176.20250   -14.28321
```

In order to check that the obtained Matrix A^{-1} was accurate, a verification check was carried out. Mat A × Mat A^{-1} was carried out to check for the obtainment of an identity matrix. An identity matrix consists of a matrix where only the diagonal line running from the top left corner to the bottom right corner has 1's in it and all other cells have 0's in them.

```
Verification Test
1.000 -0.002  0.005 -0.004  0.004 -0.004  0.002 -0.003 -0.002 -0.003 -0.003
0.000  0.998  0.005 -0.005  0.004 -0.004  0.002 -0.003 -0.002 -0.003 -0.003
0.000  0.000  1.000  0.000  0.000  0.000  0.000  0.000  0.000  0.000  0.000
0.000  0.000  0.001  0.999  0.001 -0.001  0.000  0.000  0.000  0.000  0.000
0.000  0.002 -0.004  0.004  0.997  0.003 -0.002  0.002  0.002  0.002  0.002
0.000 -0.001  0.002 -0.002  0.001  0.999  0.001 -0.001 -0.001 -0.001 -0.001
0.000  0.002 -0.004  0.004 -0.003  0.004  0.998  0.003  0.002  0.003  0.002
0.000 -0.001  0.001 -0.002  0.001 -0.001  0.001  0.999 -0.001 -0.001 -0.001
0.000  0.002 -0.005  0.005 -0.004  0.004 -0.003  0.003  1.002  0.003  0.003
0.000 -0.001  0.003 -0.002  0.002 -0.002  0.001 -0.002 -0.001  0.998 -0.001
0.000  0.000  0.001 -0.001  0.001 -0.001  0.001 -0.001  0.000 -0.001  0.999
```

Since the matrix obtained roughly corresponded to an identity matrix, it was verified that the obtained inverse matrix was accurate. This inverse matrix was then multiplied into Matrix Y. The product gave us Matrix X. The coefficients obtained in Matrix X were used to create 'predicted' values of Flux using the same time values from *Table 2* and Equation (11) from page 9 and 11 respectively. These were then graphed using bar charts to compare them to the actual values of flux from the data in *Table 2*.

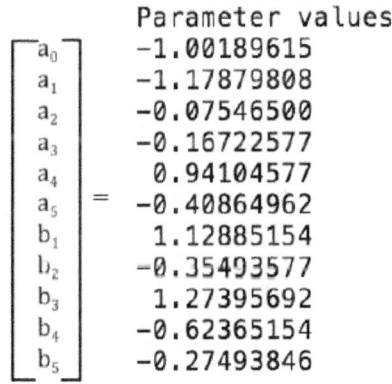

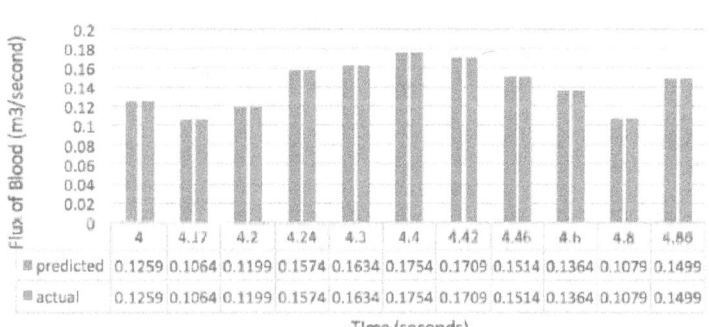

Graph 3. Aortic Flux with time: Comparison of Actual and Predicted Values

Graph 3 shown above suggest that Matrix Y was accurately obtained from the multiplication of Matrix A and Matrix X. This, of course, is expected from the solution of simultaneous equations. To test the replicability of this equation, a random set of time values were chosen from the data set that were not used in the creation of Matrix A was used to calculate flux values using Equation (11) and the parameters found from solving for Matrix X. Then, the values obtained were compared to the actual values from the data set as shown in *Graph 4*. Finally, the actual and predicted curves were plotted together in *Graph 5*.

Graph 4. Aortic Flux with time: Comparison of Actual and Predicted Values

Graph 5. Aortic Flux with time: Actual and Predicted Curves

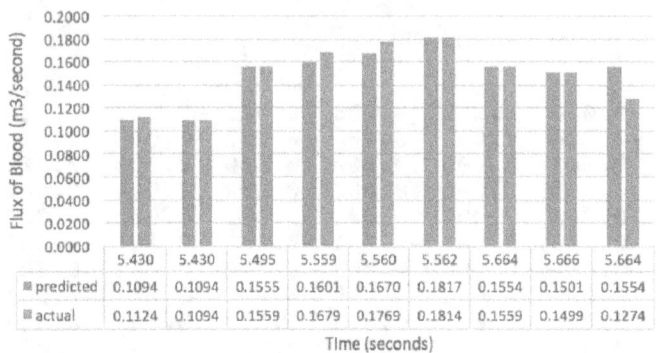
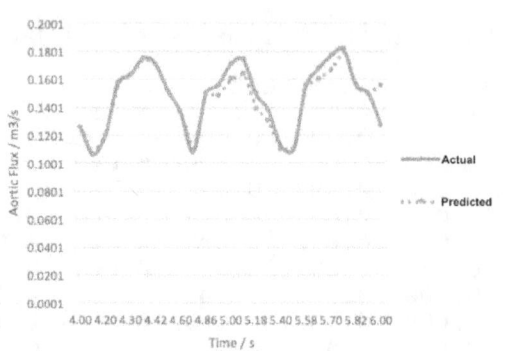

While the predicted values appeared to show slight deviations from the actual values, their degree of accuracy was to be quantitatively estimated. For this purpose, the Chi-squared test, a statistical test used to find the goodness of fit between 'observed' values and 'expected' values was used.

Null hypothesis / H_0 : There is no significant difference between the actual and predicted values of flux.

Alternate hypothesis / H_A : There is a significant difference between the actual and predicted values of flux.

Test statistic: $\chi^2 = \sum \frac{(Observed - Expected)^2}{Expected}$, where *Observed* represents the predicted or calculated values and the *Expected* represents the actual values. These values can be found from the data table embedded in *Graph 4*.

Significance level: $p < 0.05$, since this is the most common significance level chosen for biostatistics. The degrees of freedom for this test is 8. Using the table containing critical values of Chi-square distribution, the critical value (c.v.) was found to be 15.51.

$$\chi^2_{calc} = \frac{(0.1094-0.1124)^2}{0.1124} + \frac{(0.1094-0.1094)^2}{0.1094} + \frac{(0.1555-0.1559)^2}{0.1559} + \frac{(0.1601-0.1679)^2}{0.1679} + \frac{(0.1670-0.1769)^2}{0.1769} +$$

$$\frac{(0.1817-0.1814)^2}{0.1814} + \frac{(0.1554-0.1559)^2}{0.1559} + \frac{(0.1501-0.1499)^2}{0.1499} + \frac{(0.1554-0.1274)^2}{0.1274}$$

χ^2_{calc} = 1.00; 1.00 < 15.51, which is the c.v., or critical value. Thus, the null hypothesis, H_0 is accepted and the alternate hypothesis, H_A is rejected. There is no significant difference between the actual and predicted values of flux.

Conclusion

Through this investigation, I aimed to create a model for the flow of blood through the central aorta. Poiseuille's Law proved useful to calculate the flow of blood in the aorta. I hypothesised that the Fourier series would be useful and accurate in modelling the cyclic change of blood flow with time. This hypothesis was conceivable because of the compound trigonometric functions we had studied in class. However, it was also prompted by the fact that I'd used a compound trigonometric function to model circadian waves, another biorhythm like blood flow, for my Extended Essay in Biology. Having previously succeeded in accurate modelling a biorhythm using trigonometric wave function, I was more confident that this would be a viable means to model blood flow as well. During my Math IA though, I was able to more deeply delve into the Fourier series, a concept I hadn't heard of previously and one that I found extremely engaging. After I finished my first draft, I also stumbled upon a paper from MIT Media Labs which used the Fourier series to model blood pressure by monitoring and modelling the frequencies of light emitted by an individual (Poh 10767)! Overall, this experience opened my eyes towards how learning from different disciplines and experiences can often come together in unexpected scenarios as well as the wondrous ways mathematics augments the study of biology.

In my blood flow model using Fourier series, the Chi-square test showed no significant difference between the predicted and actual values for a randomly selected cycle from the data set. This showed us that the model was able to make accurate predictions for the blood flow for data values from this data set. However, a limitation was that only one data set was modelled and this model will likely not hold true for other data sets since there is a natural variation in blood flow with time between people due to varying lengths and diameters of their aortas, pressures of blood, pulse rate, etc. This means that an accurate model would also be a very specific one and useful for only blood flow data from one source or person. A similar problem was highlighted by Zhang et al. who stated that the need for personalised calculation made modelling blood flow in the medical scenario difficult (6). Thus, a meaningful extension to my investigation would be to collect pressure data from primary sources instead of a secondary source to further bolster the plausibility of my model and create a computerised algorithm to reduce the hassle of personalised calculations. Moreover, in creating an algorithm, the Fourier series could be utilised upto a much larger limit of n, since I only considered values upto $n = 5$ and a matrix to more significant figures could be obtained, since mine was only upto 5 sf, limiting the closeness of the verification test to an inverse matrix. Overall, these would further increase the accuracy of the model.

An interesting finding from my investigation was that the mathematical model was very sensitive to changes in time. Initially, while fitting the randomly selected data set to the mathematical model, the actual and calculated curves were not well fitted. In trying to figure out what was causing this, I realised that even the slightest variations in time caused large variations in the calculated flux. This makes sense because the time value is part of 10 terms in the equation. To mitigate this issue, I recollected values from the data set to more significant figures and my results were more accurate. Hence, if this model is extending to other applications, a very accurate measurement of time is necessary for it to be effective. However, this shouldn't be too difficult, since data in milliseconds is available on most standard timing devices today.

Another limitation of this investigation was that it itself had no direct applications. An extension would have been to model blood flow through a diseased aorta as well, perhaps a stenotic one to mathematically compute how the blood flow varies and to create a computer program to be able recognise blood flow data and identify blood flow as a certain percentage match to either normal or stenotic blood flow. This, in coupling with a pressure sensor and a micro-computer like Raspberry Pi would be able to work as a simple blood flow and stenosis testing device.

The results of this investigation imply that such a model could be very useful in personalised medicine to monitor blood flow. Since a mathematical model for blood flow was found, this could be implemented as algorithms to create gadgets that would be able to detect and analyse patient blood flow information, serving as a simple medical device that could be used by almost anybody, since medical services aren't available constantly and sometimes relatively inaccessible to some people. This investigation made me rediscover the beauty of mathematical modelling for providing quantitative and reproducible results with many important applications.

Works cited

Dotter, Charles T., Douglas J. Roberts, and Israel Steinberg. "Aortic length: angiocardiographic measurements." *Circulation* 2.6 (1950): 916. Web. 20 Jan. 2017.

Erbel, Raimund, and Holger Eggebrecht. "Aortic dimensions and the risk of dissection." *Heart* 92.1 (2006): 137.

Eyland, Peter. "Poiseuille's Law Derivation." *Insula: Peter's Education Website*. N.p., 20 Jan. 2015. Web. 20 Jan. 2017.

"Fourier Series." *Wolfram MathWorld*. N.p., n.d. Web. 29 Jan. 2017.

Mahaffy, Joseph M. "Riemann Sums and Numerical Integration." *Math 122 - Calculus for Biology II*. San Diego State University, 31 Mar. 2004. Web. 20 Jan. 2017.

"Online Matrix Calculator." *Blue Bit. .NET Matrix Library*, n.d. Web. 29 Jan. 2017.

Poh, Ming-Zher, Daniel J. McDuff, and Rosalind W. Picard. "Non-contact, automated cardiac pulse measurements using video imaging and blind source separation." *Optics express* 18.10 (2010): 10762-10774.

Thomas, Emma. "Matrices and Determinants: Methods and Applications." *University of Surrey: Electronics and Physical Sciences*. N.p., n.d. Web. 29 Jan. 2017.

"Viscosity and Laminar Flow; Poiseuille's Law." *OpenStax CNX*. N.p., 11 Nov. 2014. Web. 20 Jan. 2017.

Zhang, Guanqun, Jin-Oh Hahn, and Ramakrishna Mukkamala. "Tube-load model parameter estimation for monitoring arterial hemodynamics." *Engineering Approaches to Study Cardiovascular Physiology: Modeling, Estimation, and Signal Processing* (2011): 20. Web. 20 Jan. 2017.

2.. Modelling the Probability of Streaks Within a Season

Introduction

International football today has been turned into one of the highest grossing sports. The amount of money circulating in the game has become immense in recent years. This transformation of the game has meant that it has become more of a results-minded business, with attractive football less of a focus compared to winning games. Inspired by a recent study of Dutch football by two researchers from the **Maastricht Economic Research Institute on Innovation and Technology**[1] and another on English football from the University of Reading[2], I decided to focus my investigation on streaks of defeats during the season.

A manager's motivations are two-fold. At the top of the league they have the motivation of winning the title, while the managers at the bottom will have the motivation of staying in the division, specially granted the prize money for doing so. Within the results of the game, I decided to focus specifically on streaks. Streaks, both positive and negative, can have a profound impact on the mentality of teams, supporters and managers alike. This brings in an element of human psychology. Successive losses do not mean that you will necessarily be worseStreaks bring in psychological impacts and, in the case of losing, makes it feel much worse. There is nothing wrong, mathematically, to loose 4 in a row in terms of points, but in terms of confidence and staying with the manager, these streaks do make a difference. Being a keen follower of football myself, I know that fans aren't necessarily interested in the aggregate outcome of the team during the season, but will be concerned with streaks, as that will reflect how they will be viewing the club at the time.

For the purpose of this investigation, streaks were considered to be a run of at least 3 consecutive games. I decided to look into losing streaks because it is those that normally determine a manager's fate, as too many defeats in a row can make a chairman or owner lose confidence that they are the right person to lead the club forward. I have observed that streaks are a very important part of the game because of the human impact that they have, and therefore thought that investigating them would be very interesting.

In this project I first investigated the what it takes to get a manager sacked and how those who did lose their job compare to the models based on my assumptions. I then focused on streaks and explored the research question of: *Find the probability of getting at least a streak of k losses from n games.* Finally, I concluded the investigation by applying my findings to a particular Premier League club, using their own data from previous seasons, and finding the probability of them having their own streaks. I anticipate that a team could apadt my model, using their own performance statistics to produce probabilities for successive loosing streaks, so that if these streaks do occur during a season, it will allow them to place blame on the manager, or to appreciate that the performance is actually mathematically probable.

Setting out the Assumptions

To begin the investigation, I proposed a hypothetical probability distribution, as is shown in the table below. Assuming that the results of a football match can be modelled as a uniform discrete random distribution, the results can be seen below. Because it is a model I decided to base the probabilities on what would be an average team's chances for success against a team of the same or similar quality. Table 1 below has x representing the number of points from each game.

---

[1] Bruinshoofd, Allard, and Bas Ter Weel. *Manager to Go? Performance Dips Reconsidered with Evidence from Dutch Football.* Rep. no. 2001-019. N.p.: Maastricht Economic Research Institute on Innovation and Technology, n.d. Print.

[2] Bell, Adrien, Chris Brooks, and Tom Markham. *The Performance of Football Club Managers: Skill or Luck?* Rep. 19-30 ed. Vol. 1. Whiteknights: ICMA Centre, U of Reading, n.d. Print.

	Win	Draw	Lose
x (Points per game)	3	1	0
P(X=x)	1/3	1/3	1/3

Table 1: Assumed probiability distribution for allocation of points

Coming up with such values for the sake of the experiment does limit what would represent accurate results. Firstly, different teams will of course have different probabilities of winning, as can be seen from the different odds that bookmakers allocate to them. Some teams will simply have players of higher caliber than others, which can be attributed to the club's budget and the amount they pay in wages. On top of this, small external factors such as playing home or away will have an impact on the game's outcome. Finally, any previous run of games that the team is off the back of is disregarded.

To begin the investigation, I applied my hypothetical distribution to produce the expected amount of points, and compared this to the real data from every Premier League club in the 2015/16 season To do so I calculated the expected mean points per game, as can be seen below.

$$E(x) = \left(\frac{1}{3} * 3\right) + \left(\frac{1}{3} * 1\right) + \left(\frac{1}{3} * 0\right) = \frac{4}{3}$$

I then took this expected points total and compared it to how the managers actually performed in the season. To do this, I got the expected points per game value of $\frac{4}{3}$ and multiplied it by the number of games that each of the managers played, refered to as the trials. I calculated the difference between the amount that was projected for them to achieve and the amount they actually achieved, and then divided it by the number of games they played to normalize the results, allowing different numbers of trials to be compared. The table showing these numbers can be found in the appendix labelled as Table 6.

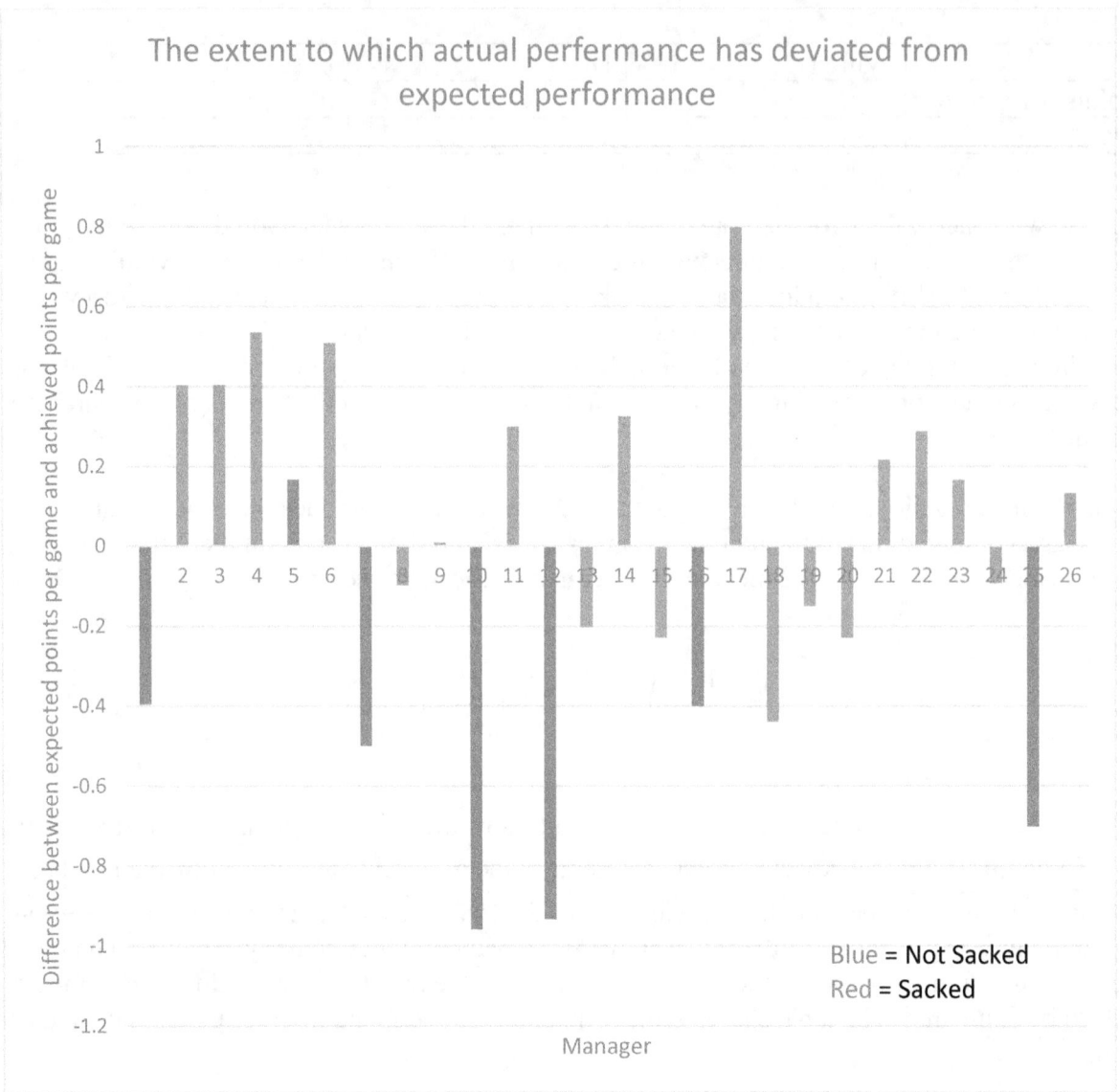

Figure 1: The difference between the expected performances based on the hypothetical distribution and actual manager performances

The bars in red show the managers who were ultimately sacked. Those after number 20 were managers hired during the season. The graph illustrates how there is a strong correlation between going below the line and managers getting fired.. This demonstrates that aggregate season performance being very influential to the sacking of managers but I noticed that there is little analysis to do with losing streaks and the impact that those have on managers keeping their jobs.

Failing to pick up points and therefore not meeting expectations comes from losing games. The decision to sack a manager is normally made off the back of recent results, so I decided to look into the probability of losing games.

I decided to simplify my model by combining the outcome of winning and drawing so that I could make a binomial analysis. I did this because, with a draw, the team still gains one point, whereas with a defeat, no points are gained. Because of this, I kept the probability of losing at 1/3 and made that of not losing 2/3. Below is the result, where x represents the number of games played.

$$P(X = x) = \binom{n}{x} * \left(\frac{1}{3}\right)^n * \left(\frac{2}{3}\right)^{n-x}$$

Although the binomial distribution is effective as a way of calculating the probability of getting a certain number of defeats within a sequence, it does not allow for the calculation of streaks because these outcomes do not have to be consecutive. I was not hapy with the limitation and went on to look for a way of solving the problem otherwise. On reflection, I decided to continue my research, as there was no suitable technique covered in the syllabus.

I found an article an analytical method produced by a physicist and a numerical one by a mathematician to the calculate the chance of getting a run of k or more results of heads from an n number of flips of a coin[3] and I was inspired by the mathematician's working and formula and decided to apply it to my own problem and make my own investigation about it.

This investigation takes the form of:
Find the probability of getting at least a streak of k losses from n games

For this investigation, the notation used for a streak of losses of k from n will be: $S(n, k)$. The requirement for this will be where n is greater than or equal to k ($n \geq k$) and the probability of a defeat (P(L)) is p. The different variables can be seen in the table below:

Variable	Meaning
S(n,k)	A streak of k or more from n
P	The probability of a defeat
L	Loss
W	Win or draw
J	Position of the first winning game

Table 2: Defining the variables

In this investigation, the events are independent of one another in order to be more simplistic

Consider different ways of getting k losses in a row

The first k games could all be losses

Result	Represented by
LLL...L	k
W...	(n-k)
All results together	n

The probability of k losses in a row is p^k

---

[3] By the Mathematician, -- By the Physicist, Combinatorics, Equations, Math, Probability. Bookmark the Permalink. "Q: What's the Chance of Getting a Run of K or More Successes (heads) in a Row in N Bernoulli Trials (coin Flips)? Why Use Approximations When the Exact Answer Is Known?" *Ask a Mathematician Ask a Physicist*. N.p., 19 July 2011. Web. 16 Aug. 2016. <http://www.askamathematician.com/2010/07/q-whats-the-chance-of-getting-a-run-of-k-successes-in-n-bernoulli-trials-why-use-approximations-when-the-exact-answer-is-known/>.

For this not to occur, a Win (w) must occur within the sequence

Assume j is in the position of the first winning game within the n games, where 1≤j≤k (can occur at the beginning of the sequence at any point until point k)

Then the probability of having k or more losses in a row in n games becomes the probability of having k or more losses in the games following the jth game

L L W * *
(W is jth match) the rest is where (j≤k), which is also n-j being that the whole thing is n

For example:
Consider the probability of getting a streak of at least 3 losses from 5 games (S(5,3)) ...

* * * * *

Consider a win occurring in the first game, (j=1)

W * * * *

Consider a win occurring in the second game, (j=2)

L W * * *

Consider a win occurring in the third game, (j=3)

L L W * *

The probability that the first Win occurs in the jth game is the probability that there are $(j-1)$ losses followed by a win

$p^{j-1} * (1-p)$ (1- probability of losing is the probability of not losing)

So for the case of j=1:

W****

the probability that the win or draw occurs in the first game is:

$$\frac{1}{3}^{1-1} * \left(1 - \frac{1}{3}\right) = \frac{2}{3}$$

In the case of j=2

LW***

the probability that the win or draw occurs in the second game is

$$\left(\frac{1}{3}\right)^{2-1} * \left(1 - \frac{1}{3}\right) = \frac{2}{9}$$

In the case of j=3

LLW**

The probability that the win or draw occurs in the third game is:

$$\left(\frac{1}{3}\right)^{3-1} * \left(1 - \frac{1}{3}\right) = \frac{2}{27}$$

The probability that the first win occurs in the jth game **and** there is a streak of k or more losses is:

$$p^{j-1} * (1-p) * S(n-j, k)$$

Where $S(n-j, k)$ refers to the sequence of games after the first win and is the probability of a streak of k or more losses happening in n-j games

So for the case of when j occurs in the first game (j=1)

W****

The probability that the win occurs in the first game, as calculated previously, **and** there is a streak of k or more losses following this is:

$$\left(\frac{2}{3}\right) * S((5-1), 3)$$
$$= \left(\frac{2}{3}\right) * S(4, 3)$$

After the win occurs in the first game, there will be four games remaining. The calculation now seeks to find the probability of a streak of three or more games within the remaining 4, expressed by S (4, 3)

For the case of when j occurs in the second game (j=2)

LW***

The probability that the win occurs in the second game **and** there is a streak of k or more losses is:

$$\left(\frac{2}{9}\right) * S((5-2), 3)$$
$$\left(\frac{2}{9}\right) * S(3, 3)$$

After the win occurs in the second game, there will be three games remaining. The calculation now seeks to find the probability of a streak of three or more games within the remaining 3, expressed by S (3, 3)

For the case of when j occurs in the third game (j=3)

LLW**

The probability that the win occurs in the third game **and** there is a streak of k or more losses is:

$$\left(\frac{2}{27}\right) * S((5-3), 3)$$
$$\left(\frac{2}{27}\right) * S(2, 3)$$

After the win occurs in the third game, there will be two games remaining. The calculation now seeks to find the probability of a streak of three or more games within the remaining 2, expressed by

S (2, 3). k has to be less than or equal to n-j, (k\<n-j), because the size of the streak cannot exceed the number of games played.

S (2,3) = 0

A streak of k or more from n number of games ($S(n, k)$) is a summation of the following probabilities:

The probability that the first k matches are losses p^k

Given that the first game is a win the probability that there is a streak of k or more losses in the remaining (n-1) games $(p^{1-1}) * (1 - p) * (S(n - 1, k))$

Given that the second game is the first win the probability that there is a streak of k or more losses in the remaining (n-2) games $(p^{2-1}) * (1 - p) * (S(n - 2, k))$

This continues until the jth game where j=k

$$S(n, k) = p^k + \sum_{j=1}^{k} p^{(j-1)} * (1 - p) * S((n - j), k)$$

[4]

Returning to the example:

So far we have the following cases:
LLL**
The probability that the sequence starts with a streak for three losses (k) is given by:
$p^k = (\frac{1}{3})^3 = \frac{1}{27}$

W****
The probability that the sequence starts with a win (j=1) and that a streak of three or more wins is achieved in the remaining games:
$\frac{2}{3} * S(4,3)$

LW***
The probability that a win occurs in the second game (j=2) and that a streak of three wins is achieved in the remaining games:
$\frac{2}{9} * S(3,3)$

LLW**
The probability that a win occurs in the third game (j=3) and that a streak of three is achieved in the remaining games:

---

[4] By the Mathematician, -- By the Physicist, Combinatorics, Equations, Math, Probability. Bookmark the Permalink. "Q: What's the Chance of Getting a Run of K or More Successes (heads) in a Row in N Bernoulli Trials (coin Flips)? Why Use Approximations When the Exact Answer Is Known?" *Ask a Mathematician Ask a Physicist*. N.p., 19 July 2011. Web. 16 Aug. 2016. <http://www.askamathematician.com/2010/07/q-whats-the-chance-of-getting-a-run-of-k-successes-in-n-bernoulli-trials-why-use-approximations-when-the-exact-answer-is-known/>.

$$\frac{2}{27} * S(2,3)$$

For the purpose of this investigation, 2 defeats in a row from is not taken into consideration as a streak, as we are only looking at 3 or more games. This is a recursive process. $S(n, k)$ is calculated by finding values for $S(n - j, k)$ for $1 \leq j \leq k$.

$S(n - j, k)$ itself may then have to be determined in the same manner. More difficult values of n, higher numbers, are replaced by lower values of n, easier values to calculate. This continues until we arrive at a self evident case, $S(a, b)$. If $a = b = k$ then $S(a, b) = p^k$. If $a < b$, for example, $S(2,3)$, then no streaks are possible. Therefore, $S(a, b)$ where $a < b = 0$

Returning to the example concerning finding the probability of a streak of 3 or more from 5 cases, we now have to calculate what the probability of getting a streak of three or more losses from the remaining cases $(S(n - j, k))$ for all of our relevant j values.

For (ii) the process continues to find S(4,3):

Again, take the case of j=1

W***

The probability that the win occurs in the first game and there there is a streak of k or more losses in the remaining games is:

$$\left(\frac{1}{3}\right)^{1-1} * \left(1 - \frac{1}{3}\right) * S((4 - 1), 3)$$

$$= \frac{2}{3} * S(3,3)$$

S(3,3) is one of the self evident cases, where a=b. The probability of getting a streak of 3 or more loses from three games is: $\left(\frac{1}{3}\right)^3$

Take the case of j=2
LW**

This is another example of a self evident case where a<b. A streak of 3 is impossible from 2 games.

For (iii) the process continues to find S(3,3). This has already been calculated from the previous case:

$$S(3,3) = \left(\frac{1}{3}\right)^3$$

For iv) the process continues to find S(2,3). This is a self evident case:

S(2,3) =0

Putting all of this together:

$$S(5,3) = \left(\frac{1}{3}\right)^3 + \sum_{j=1}^{3} \left(\frac{1}{3}\right)^{j-1} * \left(1 - \left(\frac{1}{3}\right)\right) * S(5-j, 3)$$

$$= \left(\frac{1}{3}\right)^3 + \frac{2}{3} * S(4,3) + \frac{2}{9} * S(3,3) + \frac{2}{27} * S(2,3)$$

$$= \left(\frac{1}{3}\right)^3 + \frac{2}{3}\left(\left(\frac{1}{3}\right)^3 + \left(\frac{2}{3}\right) * S(3,3)\right) + \frac{2}{9}\left(\frac{1}{3}\right)^3$$

$$= \left(\frac{1}{3}\right)^3 + \frac{2}{3}\left(\left(\frac{1}{3}\right)^3 + \left(\frac{2}{3}\right)\left(\frac{1}{3}\right)^3\right) + \left(\frac{2}{9}\right)\left(\frac{1}{3}\right)^3$$

$$= \frac{1}{27} + \left(\frac{2}{3}\right)\left(\left(\frac{1}{27}\right) + \left(\frac{2}{81}\right)\right) + \frac{2}{243}$$

$$= \frac{21}{243}$$

$$= \frac{7}{81}$$

Therefore, the probability of getting a streak of 3 or more losses from 5 games is 7/81.

To calculate the probability getting a streak of 3 or more losses in a season of 38 games is very time consuming to be written by hand. A computer program can be written to find S(38,3) or it can be taken using experimental means. This approach, known as the Monte Carlo method, was developed for "obtaining numerical solutions to problems which are too complicated to solve analytically"[5].

Monte Carlo Simulation

The random number generator on Microsoft Excel to simulate the results of a 38 game season. A number is randomly computed between 0 and 1. If the number is less than 1/3, the probability assigned for a loss, then the program produced a 0, therefore if the number is greater than 1/3, the program produces a 1. Simulating the results from an entire season would then consist of 38 trials.

The simulation was run 20 times in order to give an idea of how the probabilities look like if the number of trials increases, and the following results were obtained:

k	Monte Carlo S(38,k)	Analytical S(38,k)	Percentage difference
3	14/20 (70%)	62.72%	7.28%
4	6/20 (30%)	26.13%	3.87%
5	3/20 (15%)	9.15%	5.85%
6	1/20 (5%)	3.03%	1.97%
7	1/20 (5%)	0.99%	4.01%

Table 3: The Monte Carlo Simulation

The results were compared with the computationally generated version using the same analytical method that was done by hand previously[6]. These results were given by an excel generated

---

[5] "Monte Carlo Method." -- from *Wolfram MathWorld*. Wolfram MathWorld, n.d. Web. 16 Sept. 2016. <http://mathworld.wolfram.com/MonteCarloMethod.html>.

[6] http://maxgriffin.net/CalcStreaks.shtml

program. The numbers were randomly generated by the said program, a screen shot of it in operation can be seen in the appendix section. Excel worked by randomly generating a number between 0 and 1. If the number was below the input probability of 1/3, then it was shown as a 0, if it was above this then came came up as 1. The column next to it counted the number of defeats that occurred in a row and therefore showed the streaks. As can be seen from all the percentage differences being less than 10%, the simulation gave results very close to what would have been calculated had it been plugged in to the formula.

I looked at streaks because sometimes people make decisions based on short term decisions, but even the number that was given in the end does not illustrate the whole picture because different clubs will have different expectations and therefore the probability of defeat would be different between them.

Application

In order to take the investigation one step further, I decided to look into using probabilities based on team's performances in real life. To do so, I chose to look at Stoke City. They currently have the second longest-serving manager in the Premier League and have been a consistent side over the past three seasons where Mark Hughes has been in charge.

To make the equation more personal to them, I created new probabilities based on the results they obtained in the past three seasons, as can be seen in the table below

Season	Wins	Draws	Defeats
2013/14	13	11	14
2014/15	15	9	14
2015/16	14	9	15
Average	14	9.6666	14.3333
Average Divided by 38 games	$\frac{7}{19}$	$\frac{29}{114}$	$\frac{43}{114}$

Table 4: Stoke City's average results in the past 3 years

The average was divided by the number of games played in order to get the probability of each result per game. With these numbers, we can calculate the amount of points we would expect them to get per game by multiplying each result with the amount of points that each result would get them.

$$E(x) = \left(\frac{7}{19} * 3\right) + \left(\frac{29}{114} * 1\right) + \left(\frac{43}{114} * 0\right) = \frac{155}{114}$$

As can be seen, the expected average number of points for Stoke City to get per game would be $\frac{155}{114}$, which is just higher than the initial assumption of $\frac{4}{3}$. Multiplying that by the 38 games in a season, it could be expected for the team to get an average of 51.666 points. This amount of points could see a team finish around the middle of the table, where Stoke City would be expected to finish at the start of a season.

While Stoke may be a team that finish consistently away from the danger of relegation, a bad streak of defeats may leave them questioning whether or not to stick with Mark Hughes as their manager. Even though the manager's performance might be acceptable, the possibility of a loosing streak might trigger a mid-season firing. Again, looking only at defeats for the purpose of this investigation, the probability for Stoke to lose a game is $\frac{43}{114}$, while the probability of not losing is the remaining $\frac{71}{114}$.

These numbers for Stoke City can be applied to the same concept as the initial assumptions of each result having the probability of $\frac{1}{3}$. If the new numbers are applied using the analytical solution given by the program previously used[7], the results can be generated, as is shown on the table below.

Length of streak	Probability of streak
3 or more	75.08%
4 or more	37.81%
5 or more	15.52%
6 or more	5.91%

Table 5: The probability for Stoke City to go on different lengthed losing streaks

These probabilities could allow Stoke City as a football club to examine their situation with a more accurate scope if any of the streaks do occur. A three game losing streak may make the fans uneasy, but, when compared to the average amount of defeats they normally have per season, it could be said that they should expect to go on such a streak of results, as the probability of them doing so it 75.08%. The case would be different, however, if they were to go on a losing streak of 5 games. This only has a 15.52% chance of happening compared to their own previous results from the last three seasons, so a scenario such as this could have been down to the manager, so an evaluation of his position would be necessary.

These results offer a good evaluation of the results Stoke have obtained in the last three years. In each of the past three seasons, the club has gone on 2 separate streaks of 3 consecutive defeats. Each one of these would have a probability of 75.08% considering the overall data analysed, and therefore can be understood by fans and the decision makers at the club, who would not have worries with looking at such figures.

Conclusion

I felt like this investigation was challenging, but allowed me to go into numerical depths concerning football that I would otherwise not have been able to explore. Being an interested follower of football myself, I now how the effect of a losing streak can make clubs and fans change their opinions on players and managers, and have had to live through it with my own club, as well as watching the process unfold in several others. This project gave me the opportunity to analyse this concept and

Although happy with the results I achieved and the way I set out the project, I am very aware of its limitations and shortcomings. Firstly, as explained in its section, the hypothetical probabilities used to calculate the streaks were very basic. I chose to make it this way because I would then be able to have an easier time applying the actual mathematics of the problem using the simpler numbers, instead of being confused with more realistic numbers that would be harder to look at. Setting the probabilities of drawing winning and losing all at $\frac{1}{3}$ meant that many essential elements were not taken into account when looking at how teams would actually predict their porbabilities to be. On top of this, making each result mutually exclusive to the other meant that the experiment lost some of the human element to it. When a team is on a losing streak, the loss of confidence from the fans and the board of the club will almost always be reflected by performances on the pitch, making them more likely to carry on losing again.

---

[7] "Max Griffin--author." *Max Griffin--author.* N.p., n.d. Web. 16 Sept. 2016. <http://maxgriffin.net/CalcStreaks.shtml>.

When looking at the Monte Carlo simulation, I could have expanded my trials on excel by looking at a number higher than 20 trials. The purpose of including the simulation in the project was to show how there are different numerical methods of getting to the same result. By conducting more tests, I would have been able to get a closer result to the actual numbers I was looking for, but I felt like 20 trials gave me the means of effectively demonstrating my point.

In terms of looking at the probability of Stoke's losing runs, I could have expanded on it by looking at the possibility of more than one streak of defeats in a season. Multiple streaks close to each other can give an even more gloom look to a manager's results, while, if separate, can give the indication that whatever was changed after the first one was not a long term solution. These were present in the actual data, and I could have investigated as to how I could calculate the probability of such events taking place. On top of this, I could have done the calculation of the probabilities of the streaks by hand as a way to further challenge myself.

Bibliography

Bell, Adrien, Chris Brooks, and Tom Markham. The Performance of Football Club Managers: Skill or Luck? Rep. 19-30 ed. Vol. 1. Whiteknights: ICMA Centre, U of Reading, n.d. Print.

Bruinshoofd, Allard, and Bas Ter Weel. Manager to Go? Performance Dips Reconsidered with Evidence from Dutch Football. Rep. no. 2001-019. N.p.: Maastricht Economic Research Institute on Innovation and Technology, n.d. Print.

By the Mathematician, -- By the Physicist, Combinatorics, Equations, Math, Probability. Bookmark the Permalink. "Q: What's the Chance of Getting a Run of K or More Successes (heads) in a Row in N Bernoulli Trials (coin Flips)? Why Use Approximations When the Exact Answer Is Known?" Ask a Mathematician Ask a Physicist. N.p., 19 July 2011. Web. 16 Aug. 2016. <http://www.askamathematician.com/2010/07/q-whats-the-chance-of-getting-a-run-of-k-successes-in-n-bernoulli-trials-why-use-approximations-when-the-exact-answer-is-known/>.

"Monte Carlo Method." -- from Wolfram MathWorld. Wolfram MathWorld, n.d. Web. 16 Sept. 2016. <http://mathworld.wolfram.com/MonteCarloMethod.html>.

"Max Griffin--author." Max Griffin--author. N.p., n.d. Web. 16 Sept. 2016. <http://maxgriffin.net/CalcStreaks.shtml>.

3. Investigating the Koch's Snowflake

Introduction:

The objective of this math exploration is to investigate the significance of Koch's Snowflake. It is evident that in order to understand Swedish mathematician, Swede Niels Fabian Hedge von Koch's snowflake which was initially composed by infinite iterations of the Koch curve in 1904 (Fung), one must grasp the mathematical concepts of fractals ("Data genetics") - the creators of some of the most beautiful shapes found in mathematics that are formed through a series of infinitely small repeated patterns that make up a single geometric object. Hence, Koch's snowflake is essentially an equilateral triangle with smaller triangles added to the single three-sided shape at each segment of its edge. The investigation is of interest to me because it refers to an object which has the paradoxical property of having a finite area bounded by an infinite perimeter. Thus, being a Higher level art student, my interest in the topic of a snowflake's formation was initiated through my curiosity in the mathematics behind natural geometry. It is through understanding Koch's snowflake, which may be looked at a realistic display of an actual snowflake's composition, that underlining patterns within nature can be identified and thus, progression in human applications of computer graphics to replicate the natural world can occur (Plegner).

Rationale:

Through this investigation, I will calculate and record the process of exploring the four components of the Koch's snowflake: the number of sides, lengths of sides, perimeter, and area of the snowflake over the course of the first 5 iterations. An iteration is a repetition of an operation often involving the process of taking an output of a function and plugging it back in (Nichols). By doing so, a relationship will be identified between the values of N_n (Number of sides at an nth term), S_n (Length of a single side at an nth term), P_n (Total Perimeter at an nth term), A_n (Total Area at an nth term). This study explores the idea of a geometric object that has an infinite perimeter but finite area. It also uses these objects as a way of visualizing the limit of a geometric series.

Construction of Koch Island:

A similar object to the Koch curve is the Koch island {or Koch snowflake} which can be formed by fitting together 3 suitably rotated Koch curves. Essentially, Koch's curve which can be simplified into a 3 step process; draw a horizontal line of unit length, divide the line into three equal segments, remove the middle segment of the line and replace it with two sides of an equilateral triangle resulting in an outline of a hexagram ("Data Genetics"). Now what is left is a shape comprised of four equal length segments. Thus, the remainder of the equilateral triangle once the base is removed (Joel) is what is known to be the first iteration of the Koch's snowflake. As the process advances, it is evident that three Koch curves make a Koch snowflake.

The Koch curve is an example of a curve which is made out of corners everywhere! This means that there is no way to fit a tangent to any of its point. Since differentiation is the process of finding the gradient of a tangent to any point on a curve, this cannot be accomplished in this case. Thus, the Koch curve is a non-differentiable function.

Step 1:

Length = 1

⌞—————⌟

Step 2:

Length of Line segments = 1/3 each

⌞——⌞——⌞——⌟

Step 3:

Length of Line segments = 1/3 each

Total Length of Koch's edge = 4/3

Fractals, like the Koch curve, are formed through a construction process. At any finite stage of the construction an object is produced ("Fractals: Cantor Set"). This object is yet to become a fractal as the true fractal exists only as an idealization and can be conceived as a limit object. Thus, the construction process produces an object which approaches this limit object. This is analogous to the sum of an infinite geometric sequence, which occurs when the ratio of consecutive terms is found between -1 and 1. As terms approach or converge to a specific value, as do the sum of the terms. Hence, the limit object is the object produced by an infinite number of iterations of the original object.

Let:

N_n= The number of sides at the nth term
S_n= The length of a single side at the nth term
L_n = The total Length of the Koch Edge at the nth term
P_n= The length of the perimeter at the nth term
A_n = The area of the snowflake at the nth term

Number of Sides:

At each iteration, the number of sides of the Koch's snowflake increase at an exponential rate by a factor of 4 (each side of the snowflake from the previous iteration becomes 4 sides in the following iteration). The number of sides in the Koch's snowflake after each nth term is given by the formula:

$$N_n = 3 \times 4^{n-1}$$

Sample Calculations:

Original Object - when 'n' is equal to 1:
$$N_1 = 3 \times 4^{1-1}$$
$$N_1 = 3 \times 1$$
$$\therefore N_1 = 3$$

Object 2 (1st Iteration) - when 'n' is equal to 2:
$$N_2 = 3 \times 4^{2-1}$$
$$N_2 = 3 \times 4$$
$$\therefore N_2 = 12$$

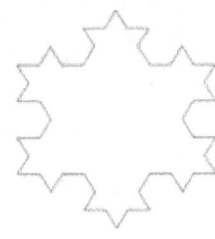

Object 3 (2nd Iteration) - when 'n' is equal to 3:
$$N_3 = 3 \times 4^{3-1}$$
$$N_3 = 3 \times 16$$
$$\therefore N_3 = 48$$

Object 4 (3rd Iteration) - when 'n' is equal to 4:

$$N_4 = 3 \times 4^{4-1}$$
$$N_4 = 3 \times 64$$
$$\therefore N_4 = 192$$

Object 5 (4th Iteration) - when 'n' is equal to 5:

$$N_5 = 3 \times 4^{5-1}$$

$$N_5 = 3 \times 256$$

$$\therefore N_5 = 768$$

6th Iteration - when 'n' is equal to 6:
$N_6 = 3 \times 4^{6-1}$

$N_6 = 3 \times 1024$

$\therefore N_6 = 3072$

Length of Single Side:

At every following iteration, there are 4 times as many line segments, each 1/3 the length of the previous iteration. This is clearly illustrated in the diagrams above.

Sample Calculations:

Original Object -
By definition, the length of the original object is unit length.

Object 2 (1st Iteration) -
$S_2 = \dfrac{1}{3} \times 1$

$S_2 = \frac{1}{3}$

∴ Each side length: $\frac{1}{3}$

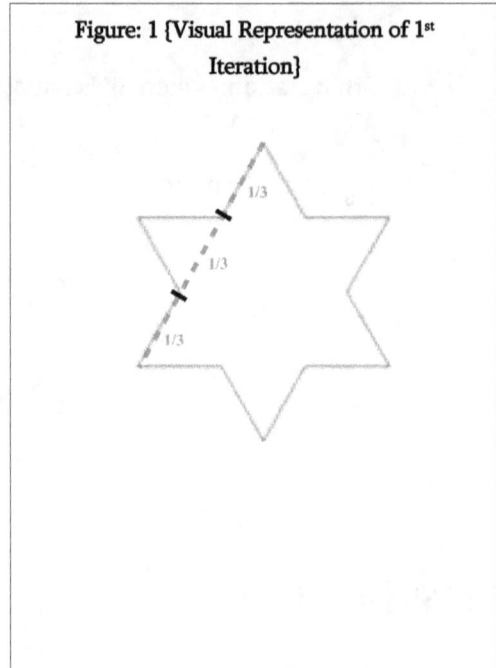

Figure: 1 {Visual Representation of 1st Iteration}

Object 3 (2nd Iteration) -

$S_3 = \frac{1}{3} \times \frac{1}{3}$

$S_3 = \frac{1}{9}$

∴ Each side length: $\frac{1}{9}$

Object 4 (3rd Iteration) -

$S_4 = \frac{1}{3} \times \frac{1}{9}$

$S_4 = \frac{1}{27}$

∴ Each side length: $\frac{1}{27}$

Object 5 (4th Iteration) -

$S_5 = \frac{1}{3} \times \frac{1}{27}$

$S_5 = \frac{1}{81}$

∴ Each side length: $\frac{1}{81}$

Object 6 (5th Iteration) -

$S_6 = \frac{1}{3} \times \frac{1}{81}$

$S_6 = \frac{1}{243}$

∴ Each side length: $\frac{1}{243}$

Since the original equilateral triangle has a side length of unit length, the length of the side of the snowflake at each 'nth' iteration can be expressed by the formula:

$$S_n = \frac{1}{3}^{n-1}$$

The reason for this being that each consecutive object's single side length is 1/3 of that of the previous object's.

Total Length of the Koch Edge:

Knowing that at each iteration there are 4 times as many line segments that are each a third the length of the previous ones, what is the formula for the total length of the Koch Edge per n number of iterations?

Koch's snowflake is "self-similar", meaning the object appears to be approximately the same on any scale (Weisstein). As the curve is magnified at any stage of iteration, it is apparent that the same detail from the first iteration is illustrated. Since at each progressive iteration the length of the line is 4/3 times the previous one, although the Koch's curve has a finite area and never intersects itself, its length is infinite. Therefore, at each iteration the length of the edge is simply multiplied by 4/3. This is represented by the formula for total length of the Koch's curve per nth iteration when L is equal to the total length:

$$L_n = \frac{4^{n-1}}{3}$$

Sample Calculations:

Original Object - when 'n' is equal to 1:

$$L_1 = \frac{4^{1-1}}{3}$$

$$\therefore L_1 = 1$$

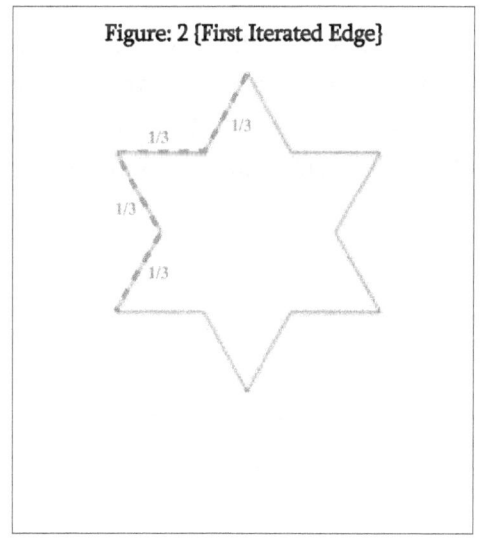

Figure: 2 {First Iterated Edge}

Object 2 (1st Iteration) - when 'n' is equal to 2:

$$L_2 = \frac{4^{2-1}}{3}$$

$$\therefore L_2 = \frac{4}{3}$$

Object 3 (2nd Iteration) - when 'n' is equal to 3:

$$L_3 = \frac{4^{3-1}}{3}$$

$$\therefore L_3 = \frac{16}{9}$$

Object 4 (3rd Iteration) - when 'n' is equal to 4:

$$L_4 = \frac{4^{4-1}}{3}$$
$$\therefore L_4 = \frac{64}{27}$$

Object 5 (4th Iteration) - when 'n' is equal to 5:

$$L_5 = \frac{4^{5-1}}{3}$$
$$\therefore L_5 = \frac{256}{81}$$

Object 6 (5th Iteration) - when 'n' is equal to 6:

$$L_6 = \frac{4^{6-1}}{3}$$
$$\therefore L_6 = \frac{1024}{243}$$

Hence, the length of the curve per nth term is $\frac{4^{n-1}}{3}$ times the previous object. Since n increases infinitely however, the length of the curve increases with no bound, resulting in an infinite perimeter which can be expressed through the equation below. Here, irrespective of an increase in the value of n, the total perimeter of the snowflake remains relatively constant.

Total Perimeter:

Since all the sides of Koch's snowflake are equal in length per nth iteration, and perimeter can be defined as the sum of the lengths of the sides of a shape, it can be calculated by multiplying the number of sides of the snowflake by the length of each side at each stage. Therefore, the formula for the total length of the snowflake's outline is simply:

$$P_n = N_n \times S_n$$

$$\therefore P_n = (3 \times 4^{n-1})\left(\frac{1}{3}^{n-1}\right)$$

Sample Calculations:

Original Object: when 'n' is equal to 1:
$P_1 = 3 \times 1$

$\therefore P_1 = 3$

1st Iteration: when 'n' is equal to 2:

$P_2 = 12 \times (1/3)$

$\therefore P_2 = 4$

2nd Iteration: when 'n' is equal to 3:

$P_3 = 48 \times (1/9)$

$\therefore P_3 = 16/3$

3rd Iteration: when 'n' is equal to 4:

$P_4 = 192 \times (1/27)$

$\therefore P_4 = 64/9$

4th Iteration: when 'n' is equal to 5:

$P_5 = 768 \times (1/81)$

$\therefore P_5 = 256/27$

5th Iteration: when 'n' is equal to 6:

$P_6 = 3072 \times (1/243)$

$\therefore P_6 = 1024/81$

The sample calculations up to the 5th iteration of the Koch's curve are presented in the table below:

n	N_n	S_n	P_n
1	3	1	3
2	12	$\frac{1}{3}$	4
3	48	$\frac{1}{9}$	$\frac{16}{3}$
4	192	$\frac{1}{27}$	$\frac{64}{9}$

| 5 | 768 | $\frac{1}{81}$ | $\frac{256}{27}$ |
| 6 | 3072 | $\frac{1}{243}$ | $\frac{1024}{81}$ |

Graphs showing exponential relationships.

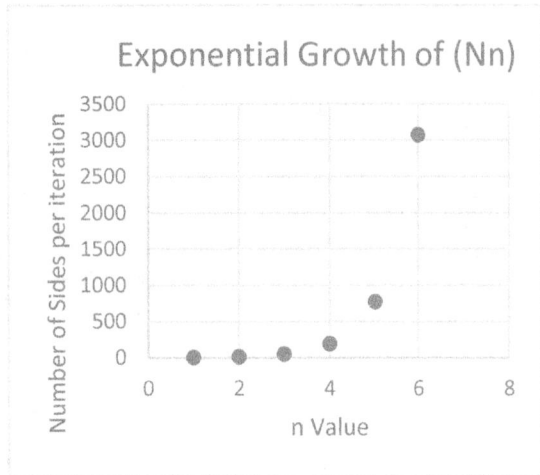

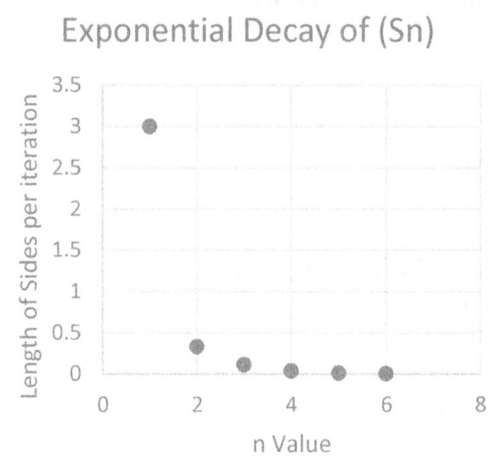

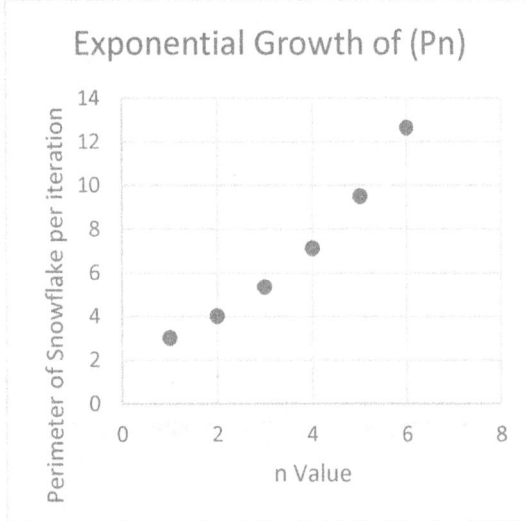

The graphs represent the exponential growth and decay of the total perimeter, area and number of sides per iteration of the Koch's snowflake. Thus, it is evident that there is a rapid change in the quantities investigated as the nth term increases and decreases.

The astronomical type of growth that can be found through exponential functions is conveyed through the following calculations:

100th Iteration: When 'n' is 101 –

N_n: $3 \times 4^{n-1}$

$N_{101} = 3 \times 4^{101-1}$

∴ $N_{101} = 4.82E60$

$S_n: \frac{1}{3}^{n-1}$

$S_{101} = \frac{1}{3}^{101-1}$

$\therefore S_{101} = 1.94\text{E}-48$

$P_n: N_n \times S_n$

$P_{101} = 4.82\text{E}60 \times 1.94\text{E}-48$

$\therefore P_{101} = 9.35\text{E}12$

Furthermore, another way of constructing the Koch island can be used in order to visualize the limit of a geometric series. At each stage of the construction process scaled down versions of the original object will be added.

Step 1: Choose an equilateral triangle T with side a
Step 2: scale down T by a factor of 1/3 and paste on 3 copies of the resulting triangle. The resulting object is bound by 3x4 straight line segments, each of length a/3
Step 3: scale down T by a factor of 1/3 x 1/3 and paste on 3x4x4 copies of the resulting triangle. The resulting object is bound by 3x4x4 straight line segments, each of length (1/3 x 1/3) a
Step 4: continue this process, resulting in the formation of the formal for the length of the individual sides in the iteration: $S_n = \frac{1}{3}^n$

Hence, in order to 'add up geometric objects' we will consider their area at successive stages of construction.

Area:

The area of the original object (T) can be solved through the use of the Area = ½ ab sin C formula. This product of this formula will be the starting point for the following areas of the Koch's snowflake in the progressive stages.

Here, the area of the first equilateral triangle T is defined by the function:

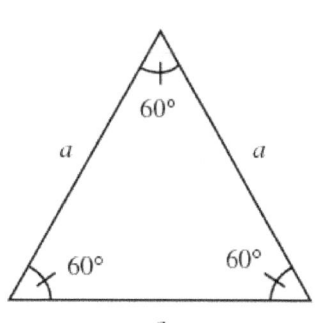

$(A_T) = \frac{1}{2} \times a^2 \times \sin(60)$

$(A_T) = \frac{\sqrt{3}}{4} a^2$

Provided with the knowledge that the next stage of the Koch's snowflake is produced by adding 3 triangles each of length $\frac{a}{3}$ to T, using set notation this can be represented as:

T ∪ 3 x (1/3 T)

Here, the iteration's length is 1/3 times the previous object. As a result, each triangle will have an area of $\frac{1}{2}$ x $\frac{a}{3}$ x $\frac{a}{3}$ x $\frac{\sqrt{3}}{2}$ = $\frac{\sqrt{3}\,a^2}{36}$

Since there are 3 of these triangles the area added will be 3 x $\frac{\sqrt{3}\,a^2}{36}$ = $\frac{\sqrt{3}\,a^2}{12}$

Hence, the area of this new object or the first iteration is calculated by:

$$(A_{T1}) = \frac{\sqrt{3}}{4}a^2 + \frac{\sqrt{3}\,a^2}{12} = \frac{\sqrt{3}}{3}a^2$$

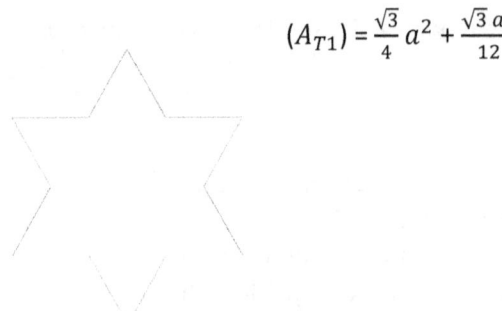

The area of the Koch's snowflake's second iteration is produced by adding 3x4 triangles each of length $\frac{a}{3}$ x $\frac{a}{3}$ to T which is essentially 12 equilateral triangles as each of the 3 sides have now been divided into 4 line segments. Therefore, using set notation this can be represented in the formula:

T ∪ 3 x (1/3 T) ∪ 12 x (1/9T)

Each triangle will have an area of $\frac{1}{2}$ x $\frac{a}{9}$ x $\frac{a}{9}$ x $\frac{\sqrt{3}}{2}$ = $\frac{\sqrt{3}\,a^2}{324}$

Thus, $\frac{\sqrt{3}\,a^2}{324}$ is the area of a single new equilateral triangle. Now knowing that at the 2nd iteration's number of sides is equal to 12, the total area of the added triangles can be calculated by multiplying 12 to $\frac{\sqrt{3}\,a^2}{324}$.

12 x $\frac{\sqrt{3}\,a^2}{324}$ = $\frac{\sqrt{3}\,a^2}{27}$

Therefore, the area of this new shape which is composed of $(A_T) + (A_{T1}) + \frac{\sqrt{3}\,a^2}{27}$, is calculated:

$$(A_{T2}) = \frac{\sqrt{3}}{3}a^2 + \frac{\sqrt{3}\,a^2}{27} = \frac{10\sqrt{3}}{27}a^2$$

Since this process is continuous, in order to find the area of A_{T3}, A_{T4} is added to the area produced by 3 x 4 x 4 triangles each of length $\frac{a}{3} \times \frac{a}{3} \times \frac{a}{3}$.

Therefore, the area of the object at different stages of iteration:

Length of Line Segments	Area of Triangle	Number of Triangles	Total Area A_{Tn}
a	$\dfrac{\sqrt{3}\,a^2}{4}$	1	$\dfrac{\sqrt{3}\,a^2}{4}$
$\dfrac{1}{3}a$	$\dfrac{\sqrt{3}\,a^2}{36}$	3	$\dfrac{\sqrt{3}\,a^2}{12}$
$\dfrac{1}{9}a$	$\dfrac{\sqrt{3}\,a^2}{324}$	12	$\dfrac{\sqrt{3}\,a^2}{27}$

Hence…

	Length of Line Segments	Area of Triangle	Number of Triangles	Total Area A_{Tn}
Total Area of Object T_n	$\left(\frac{1}{3}\right)^n a$	$\frac{\sqrt{3}}{4}\left(\frac{1}{3}\right)^{2n} a^2$	$3 \times 4^{n-1}$	$A_{Tn} = (3 \times 4^{n-1})\left(\frac{\sqrt{3}}{4}\right)\left(\frac{1}{3}\right)^{2n}(a^2)$

Table of Methodology – Finding the area of the iterated object:

Object	Image	Area of Iterated Object A_{Tn}
T	(equilateral triangle, side a, angles 60°)	$\frac{\sqrt{3}\, a^2}{4}$
$T \cup 3 \times (1/3\,T)$	(Star of David shape)	$\frac{\sqrt{3}\, a^2}{4} + \frac{\sqrt{3}\, a^2}{12}$
$T \cup 3 \times (1/3\,T) \cup 12 \times (1/9T)$	(Koch snowflake iteration)	$\frac{\sqrt{3}\, a^2}{4} + \frac{\sqrt{3}\, a^2}{12} + \frac{\sqrt{3}\, a^2}{27}$
Fractal or 'Limit Object'	N.A	$A_n = A_T + A_{T1} + A_{T2} + \cdots A_{Tn}$ $A_n = A_T + \sum_{r=1}^{n} A_{Tr}$

Geometric Sequence to Infinity:

$A_n = A_T + A_{T1} + A_{T2} + \cdots A_{Tn}$

Explanation: In order to find the total area up to the nth iteration, the sum of the area of the original triangle, 3 smaller triangles, 12 even smaller triangles and so forth (up to the nth term) must be calculated.

Therefore, a condensed way of rewriting $A_n = A_T + A_{T1} + A_{T2} + \cdots A_{Tn}$ is:

$$A_n = A_T + \sum_{r=1}^{n} A_{Tr}$$

Here, **n** is the term that indicates where you stop the process, and **r** is the individual iteration.

Considering $\sum_{r=1}^{n} A_{Tr}$, one can substitute the A_{Tr} with $3 \times 4^{r-1}) \left(\frac{\sqrt{3}}{4}\right) \left(\frac{1}{3}\right)^{2r} (a^2)$

Hence...

$$= \sum_{r=1}^{n} (3 \cdot 4^{r-1}) \left(\frac{\sqrt{3}}{4}\right) \left(\frac{1}{3}\right)^{2r} a^2$$

This can be further simplified algebraically-

Steps to the process:

Fractions rewritten using negative exponents: $\left(\frac{1}{3}\right)^{2r} = 3^{-2r}$

Numbers with same base combined and negative exponent written in fraction form:
$(3)(3^{-2r}) = \frac{3}{3^{2r}}$ & $(4^{r-1}) = \frac{4^r}{4}$

r terms separated from numbers independent from r: $\left(\frac{3}{3^{2r}}\right)\left(\frac{4^r}{4}\right) = \left(\frac{3}{4}\right)\left(\frac{4^r}{3^{2r}}\right)$

Factor out $\frac{3}{4}$: $\frac{\sqrt{3}}{4} a^2 \left(\sum_{r=1}^{n} \left(\frac{3}{4}\right)\left(\frac{4^r}{3^{2r}}\right)\right) = \left(\frac{3}{4}\right)\frac{\sqrt{3}}{4} a^2 \left(\sum_{r=1}^{n} \left(\frac{2^{2r}}{3^{2r}}\right)\right)$

Multiplication and factoring out common r term: $\left(\frac{3}{4}\right)\left(\frac{\sqrt{3}}{4}\right) = \frac{3\sqrt{3}}{16}$ & $\left(\frac{2^{2r}}{3^{2r}}\right) = \left(\frac{2^2}{3^2}\right)^r$

Calculations:

$$= \frac{\sqrt{3}}{4}a^2(\sum_{r=1}^{n}(3\cdot 4^{r-1})\left(\frac{1}{3}\right)^{2r})$$

$$= \frac{\sqrt{3}}{4}a^2(\sum_{r=1}^{n}(3)(4^{r-1})(3^{-2r}))$$

$$= \frac{\sqrt{3}}{4}a^2(\sum_{r=1}^{n}\left(\frac{3}{3^{2r}}\right)\left(\frac{4^r}{4}\right))$$

$$= \frac{\sqrt{3}}{4}a^2(\sum_{r=1}^{n}\left(\frac{3}{4}\right)\left(\frac{4}{3^{2r}}\right)^r)$$

$$= \frac{3\sqrt{3}}{16}a^2\left(\sum_{r=1}^{n}\left(\frac{2^2}{3^2}\right)^r\right)$$

$$= \frac{3\sqrt{3}}{16}a^2\left(\sum_{r=1}^{n}\left(\frac{4}{9}\right)^r\right)$$

Once $= \frac{\sqrt{3}}{4}a^2(\sum_{r=1}^{n}(3\cdot 4^{r-1})\left(\frac{1}{3}\right)^{2r})$ is reduced to its simplest form, one is able to generate the expression for the geometric sequence:

$$\sum_{r=1}^{n}\left(\frac{4}{9}\right)^r$$

For instance, the sum of the geometric sequence up to the 6th term is given below:

$$S_1 = \left(\frac{4}{9}\right)^1$$

$$S_2 = \left(\frac{4}{9}\right)^1 + \left(\frac{4}{9}\right)^2$$

$$S_3 = \left(\frac{4}{9}\right)^1 + \left(\frac{4}{9}\right)^2 + \left(\frac{4}{9}\right)^3$$

$$S_4 = \left(\frac{4}{9}\right)^1 + \left(\frac{4}{9}\right)^2 + \left(\frac{4}{9}\right)^3 + \left(\frac{4}{9}\right)^4$$

$$S_5 = \left(\frac{4}{9}\right)^1 + \left(\frac{4}{9}\right)^2 + \left(\frac{4}{9}\right)^3 + \left(\frac{4}{9}\right)^4 + \left(\frac{4}{9}\right)^5$$

$$S_6 = \left(\frac{4}{9}\right)^1 + \left(\frac{4}{9}\right)^2 + \left(\frac{4}{9}\right)^3 + \left(\frac{4}{9}\right)^4 + \left(\frac{4}{9}\right)^5 + \left(\frac{4}{9}\right)^6$$

It is evident from the previous calculations in reference to the area of the Koch snowflake that the first term of the sequence or u_1 is $\left(\frac{4}{9}\right)^{1-1} = 1$. In order to find the common ratio of the sequence of areas, the change A_{T1} and A_{T2} is calculated where 'r' = $\left(\frac{U1}{U2}\right)$

Hence, the common ration, 'r', is $\frac{4}{9}$.

Since $|r| < 1$ there exists a sum to infinity $\Rightarrow \left|\frac{4}{9}\right| < 1$ (Buchanan, Laurie).

The sum to infinity of this series is given by the formula:

$$S_\infty = \frac{u_1}{1-r}$$

$S_\infty = \frac{4/9}{1-\left(\frac{4}{9}\right)} = \frac{\left(\frac{4}{9}\right)}{\left(\frac{5}{9}\right)}$

$\therefore S_\infty = \frac{4}{5}$

Previously we found that $A_T = \frac{\sqrt{3}}{4}a^2$

∴ The area of the finite or 'limit object' known as A_∞ is:

$A_\infty = A_1 + \frac{3\sqrt{3}a^2}{16}(S_\infty)$

$A_\infty = \frac{\sqrt{3}}{4}a^2 + \frac{3\sqrt{3}a^2}{16}\left(\frac{4}{5}\right)$

$A_\infty = \frac{2\sqrt{3}}{5}a^2$

This value is the area of the object as fractal, which means that it is the limit object produced by the process of construction.

Conclusion:

In essence, my objective of exploring the Koch's snowflake through calculating the object's number of sides, lengths of sides, total perimeter, and total area of the snowflake over the course of the first 5 iterations was met. Through investigating the following components (N_n, S_n, P_n, A_n), a geometric sequence & series with a common ratio of $\frac{4}{9}$ was identified. The total area of the object at each iteration was modeled by calculating the area of the individual triangles within the first 5 iterations, and thus, formulating an overarching, generalized formula by which the total area of the object per nth term (iteration) can be calculated. Moreover, since the Koch's snowflake has an infinite perimeter and finite area, the geometric sequence's sum to infinity was modeled using the sum to infinity formula; where both

u_1 and r equal $\frac{4}{9}$.

Reflection:

Evidently, although I believe this exploration was conducted in a clear, and logical manner, there are drawbacks to the process in which I investigated the Koch's Snowflake. Since von Koch's Snowflake is essentially an infinite geometric sequence, representing the snowflake in terms of the first 5 iterations is limiting. Furthermore, due to my interest in the geometric form of the fractal, a drawn study of snowflake would have contributed further to my understanding of its formation, and development of its iterations. However, having said that, I believe my knowledge of Koch's snowflake has been heightened through my personal engagement in the process of calculating its number of sides, length of sides, perimeter and area without full reliance on the fractal's given formulas. Rather that solely utilizing given formulas to guide my investigation, I first developed my own understanding of the snowflake's iterations through applying my past knowledge of sequences and series. Ultimately, the research I conducted on the Koch's snowflake supported the conclusions I made along the way, rather than resulted in them.

Works Cited

Buchanan, Laurie, Jim Fensom, Ed Kemp, Paul La Rondie, and Jill Stevens. IB Mathematics. Oxford: Oxford UP, 2014. Print.

"Fractals: Cantor Set, Sierpinski Triangle, Koch Snowflake, Fractal Dimension." *SpringerReference* (n.d.): 1-7. *Uwosh.edu*. The University of Wisconsin Oshkosh. Web. 6 Dec. 2016.

Fung, Emily. "Koch's Snowflake." Math.*ubc*. *University of British Columbia*, n.d. Web. 06 Dec. 2016.

Huang, Shaoming, and Liming Dai. "World of Fractals." *Journal of Nanoparticle Research* 4.1/2 (2002): 145-55. *Math NUS*. Web. 6 Dec. 2016.

Joel. "Koch Curve – Fractal Thought Experiment." *Algorithmic Universe*. N.p., 02 Dec. 2014. Web. 6 Dec. 2016.

"Koch Snowflake." *Data Genetics*. N.p., n.d. Web. 06 Dec. 2016.

- *Wolfram Math World*. N.p., n.d. Web. 06 Dec. 2016.

Nichols, Erin. "Applications of Iteration." Math Forum. N.p., 29 May 1997. Web. 27 Jan. 2017.

Plegner, Philip. "Fractals | World of Mathematics." *Mathigon.org*. N.p., n.d. Web. 06 Dec. 2016.

Weisstein, Eric W. "Self-Similarity." Wolfram MathWorld. N.p., n.d. Web. 28 Jan. 2017.

4. Pigeons, Birthdays, and 607 Students

Introduction

In the seventh grade, I reluctantly met another student in my Chinese class who shared my birthday. As a thirteen-year-old who had once thought that his date of birth was key in defining who he was, I was unsurprisingly very disappointed and in a mild state of disbelief.

It was not until later on in my high school career that this idea came back to me, when I came across a simple, yet perplexing mathematical proposition otherwise known as the pigeonhole principle.

The principle is simple in the aspect that it states the mere following:

> If n items are put into m containers, and $n > m$, then at least one container must contain more than one item.

However, the statement is puzzling when its vast implications are put into application within other subject matters. My seventh grade birthday dilemma is case in point, even though at first, a relationship between the two may not be entirely obvious.

Figure 1: Ben Dale, *The Pigeonhole Principle*, July 2008

To clarify using the pigeonhole principle: m would easily represent the number of days in a year, or 365 if 29 February were omitted. This is to say that there are 365 possible birthdays, or 'containers' that are available given one year.

According to the proposition, n would have to be a number larger than 365 to ensure that two people will have the same birthday. In this particular case, since the only sure case of a shared birthday was between me and the other student, it is possible to assume that n, or the number of students, would easily be 366. This would suggest that all other 364 students taking part in this hypothetical situation would have their own unique birthdays, with only me and the other student sharing the same birthday.

In layman's terms, this information would suffice to explain how one's birthday could be shared using the pigeonhole principle. To me however, this explanation merely scratches the surface of a coincidental situation that I believe has more substance to.

Aim and Rationale

In this investigation into the relationship between the pigeonhole principle and the birthday problem, I aim to calculate the probability of two people having the same date of birth given n people.

Using my calculated formula, I will expand my understanding of statistics and this particular concept by determining the probability that two people in my high school will share the same birthday.

This aim represents a question and topic that I want to explore because I think that the concept behind the pigeonhole principle is one that is simple, but has the potential to be applied to many other situations. It suggests how simple mathematics has the potential to unlock many surprising answers, and this idea intrigues me.

The Combination and Permutation Formula

To begin, I must first determine all possible combinations for somebody to have the same birthday. The following combination formula can be used to calculate this[8].

Combination Formula $C(n)$	**Permutation Formula** $P(n)$
$C(n,k) = \dfrac{P(n,k)}{k!}$	$P(n,k) = \dfrac{n!}{(n-k)!}$

where n is the size of the set from which elements are permutated
while k is the size of each permutation.

Seeing as there is a formula within the combination formula, it would be best to start there. In order to grasp a better understanding of the combination formula, I will first examine the permutation formula to see how it correlates with finding combinations and therefore the birthday problem.

ABC	ABD	ACD	BCD
ACB	ADB	ADC	BDC
BAC	BAD	CAD	CBD
BCA	BDA	CDA	CDB
CAB	DAB	DAC	DBC
CBA	DBA	DCA	DCB

Figure 2: Math and Multimedia, *Permutations (white and blue spaces) and Combinations (blue*

Permutations refer to the action of changing an arrangement, with particular reference to the order, of a set of items. Thus, permutations can be seen as lists where orders matter.

On the contrary, a **combination** is a selection of a given amount of elements from a larger number, without regard to their arrangement. In other words, combinations are merely groups, where order does not play a role in differentiation.

The Permutations Formula

Recall the permutations formula stated earlier.

$$P(n,k) = \frac{n!}{(n-k)!}$$

where n is the size of the set from which elements are permutated
and k is the size of each permutation.

I will explain this with the aid of an example. Suppose that there are four people, A, B, C, and D, in a room. In this instance, n would be the quantity of people, and thus n would be seven.

---

[8] According to the 2008 online videos *Permutations* and *Combinations* by Salman Khan of Khan Academy.

I will assume that there are only two chairs in the room. Since the chairs limit the size of our permutation, the *k* value, or the size of each permutation, is three.

If everybody in the room were first told to stand, and two individuals were asked to take a seat, then for the first seat, there would be four possible people who could have a seat. After any of the four people have taken a seat, however, the next chair would only be able to seat one of the remaining three people who are still standing.

Noting that these integers represent possibilities, I would have to multiply each with each other to determine all the different orders possible. This thus explains the concept of **factorials**, or the product of an integer and all the integers below it.

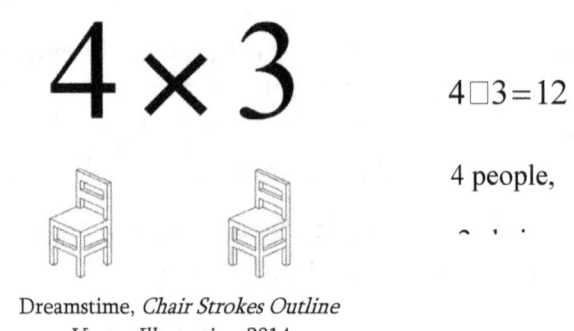

$4 \square 3 = 12$

4 people,

Dreamstime, *Chair Strokes Outline Vector Illustration*, 2014

Table 1: Possible Permutations P(4,2)

AB	BA	CA	DA
AC	BC	CB	DB
AD	BD	CD	DC

There would thus be 12 permutations with four people and two chairs.

To create a generalized equation, a similar concept can be used. For every seat taken, there is one less person who can take the next seat, which explains how the factorial aspect of the equation comes into play.

Possibilities	n	$n-1$	$n-2$	$n-3$	1
Term	1	2	3	4	n^{th}

However, the factorial for *n* ignores the size of the permutation, or *k*. In the example with seven people above, the *k* value was three, as there were only three chairs in the room.

In the first example, only the first two digits in the factorial of four were used, omitting the final two digits. Considering the entire factorial for four, it can also be seen that two factorial has been omitted.

4 Factorial	Permutations for 4 people and 2 chairs

| $4! = 4 \cdot 3 \cdot 2 \cdot 1$ | $4 \cdot 3$ |

Considering the context of the question, two can be calculated by subtracting the original k value from the n value, or two and four respectively from the example.

One way to omit two factorial from the example is to set it as the denominator, as it will cancel out with the numerator, leaving the first two digits in four factorial.

$$Permutations = \frac{4 \cdot 3 \cdot 2 \cdot 1}{2 \cdot 1} = \frac{4!}{2!}$$

Given 4 is n and 2 is k, the following equation can be calculated:

$$P(4,2) = \frac{4!}{(4-2)!}$$

$$P(n,k) = \frac{n!}{(n-k)!}$$

This thus provides an explanation for the permutations formula, which makes up an imperative part of the combination formula.

The Combination Formula

Recall that the combination formula from the information mentioned earlier.

$$C(n,k) = \frac{P(n,k)}{k!}$$

Considering the explanation to the permutations formula, the combination formula may be easier to interpret if it is rewritten.

$$C(n,k) = \frac{\frac{n!}{(n-k)!}}{k!}$$

Once again, I can use the example of four people and two chairs to this time determine the amount of combinations, noting that combinations disregard order.

For this reason, I will divide the number of permutations by the amount of ways we can arrange k items. This makes sense, because all possible permutations include repeated combinations, and in order to determine all combinations, the extraneous permutations must be removed. Refer to *Figure 2* for a visual interpretation of this idea. Since the limitation of the permutations is synonymous with the different arrangement possibilities is the combinations formula, the k value is the same.

To determine the number of ways to arrange k items into n (which in this case just represents a variable) positions, I can use the concept of factorials again. Once the first item is placed, there would be one less item left to place in the following spot. This pattern continues until the nth position, where only one item is left to be placed.

Arrangements	k	$k-1$	$k-2$	$k-3$	1
Position	1	2	3	4	n^{th}

This explains why the permutations formula plays such a large role in the combinations formula.

To simplify the formula, it may be rewritten:

$$C(n,k) = \frac{\frac{n!}{(n-k)!}}{k!}$$

$$C(n,k) = \frac{n!}{k!(n-k)!}$$

The Birthday Problem

To begin with the birthday problem, I must first determine how many possible birthday combinations exist given n people. Using the formula above, if I am looking for pairs, my k value would be represented with a two. Considering that I am solving for a general formula, n will remain n.

$$C(n,2) = \frac{n!}{2!(n-2)!}$$

Since I am determining an equation for birthdays, I must take into account the number of days in a year. I will be omitting 29 February, so there will be 365 days in a year for my generalized equation.

At this point, it is important to note that although my goal is to determine the probability of somebody sharing a birthday, judging from creating a formula, it will be easier to solve for the probability that somebody will not share my birthday. Once subtracted from a hundred percent probability, this value will allow me to determine the true probability of two people sharing the same birthday. Thus, my equation will include all the chances that two people will have different birthdays over all possible birthday outcomes. In other words, if I were to meet somebody who shared the same birthday, that day would merely represent one day out of all 365.

I will subtract this likelihood from one, as this represents a hundred percent probability. Now, an important clarification should be made about percentages and probability. Whereas a percentage is a probability determined out of a hundred, probability merely measures the extent to which something may occur out of one.

$$\frac{365}{365} - \frac{1}{365} = \frac{364}{365}$$

$$= 0.997$$
$$= 99.7\%$$

Note: calculations will be rounded to three significant figures throughout, as this will provide an appropriate degree of accuracy.

As I will be determining this probability for multiple combinations, I can use an exponent to ensure that my calculated probability for two people not sharing a birthday is multiplied by all possible combinations. The amount of combinations for *n* people has already been calculated above, so this will become the exponent for the probability. Assuming that birthdays are independent, this means the probability will be multiplied by itself as many times as there are combinations available (see below).

Probability of not sharing a birthday where *n* represents the quantity of people being tested for.	$\left(\dfrac{364}{365}\right)^{\frac{n!}{2!(n-2)!}}$

To determine the probability of sharing the same birthday, I merely need to subtract the probability of not sharing the same birthday from all total possibilities. Thus, my calculated formula would change just a little.

Probability of sharing a birthday where *n* represents the quantity of people being tested for.	$P(n) = 1 - \left(\dfrac{364}{365}\right)^{\frac{n!}{2!(n-2)!}}$

Now that my finalized equation has been determined, for any value of *n*, I can now determine the probability of two people sharing the same birthday. Below, I have included a table that determines the probability for different values of *n*.

A sample calculation up to three significant figures is included for one value of *n*.

$$P(n) = 1 - \left(\dfrac{364}{365}\right)^{\frac{n!}{2!(n-2)!}}$$

$$P(10) = 1 - \left(\dfrac{364}{365}\right)^{\frac{10!}{2!(10-2)!}}$$

$$P(10) = 1 - 0.884$$

$$P(10) = 0.116$$

Probability, given 10 people, in percentage (sample)	11.6

Below are variable values for *n*, including their associated probabilities:

Table 2: Calculated Probability for Variable Values of n

n	P(n)	Probability (%)
10	0.116	11.6
20	0.406	40.6
23	0.500	50.0
30	0.697	69.7
40	0.882	88.2
50	0.965	96.5
60	0.992	99.2

Note: highlighted column to be utilized for reference later.

Determining the Probability of a Shared Birthday

To visually interpret this relationship, I have used an online software:

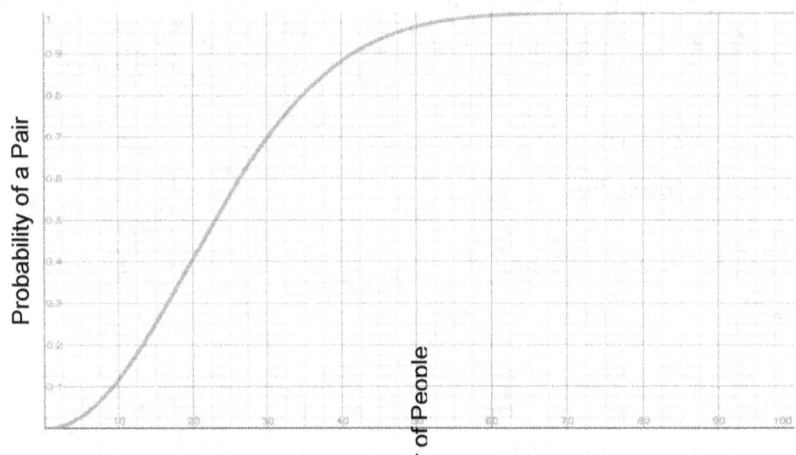

Figure 3: Graph of Birthday Probabilities (Graphed using online software Desmos)

To apply my understanding into a real-life situation, I decided to test this phenomenon in my school. As part of my aim, I will attempt to determine the probability that two people in my high school will share the same birthday. In particular, I would like the find the probability of a birthday pair in classes of 23 people.

I decided to use groups of 23 because my calculated probability when *n* equals 23 is 50 percent (rounded). Refer to *Table 2* or *Figure 3*. This represents a high probability given a small integer *n*, and thus it would be interesting to apply this to my situation.

An important consideration is my methodology into this investigation. The following points illustrate the most important aspects of my approach:

Data source – Procuring the birthdays of all high school students required me to reach out to my school's high school office, where school data is reliable and accurate. Personal student details were not disclosed during this process. Refer to appendix for complete list of birthdays.

Use of an online mechanism – In order to classify all 607 students into groups of 23, the use of technology in the form of an online 'group generator' was used. As 607 is indivisible by 23, several birthdays were not included in trials. To be precise, for each of five trials, 25 classes were generated each containing 23 students, meaning that only 575 birthdays were utilized in one trial and 28 birthdays were omitted per trial.

Data processing – In the event that more than one pair of shared birthdays occurred in one class, the group was only considered for having one birthday pair, as my objective aims to find the probability of one pair given 23 people.

Use of five trials – A **mean** must be calculated in the end as multiple trials are utilized to increase accuracy. Multiple trials are a statistical consideration that aims to bring theoretical and experimental probability closer together, and the mean determine the sum of all quantities over their number or quantity.

Assuming birthdays are independent – Birthdays are assumed to be independent and not dependent on given situations, even though all data comes from the school.

Below includes a raw data table for all total birthday pairs for each trial and class:

Table 3: Occurrence of Total Birthday Pairings in All Classes

Class	Quantity of Birthday Pairings				
	Trial 1	Trial 2	Trial 3	Trial 4	Trial 5
1	1	0	2	0	0
2	0	0	0	0	0
3	1	1	1	0	1
4	0	1	0	0	0
5	1	0	0	0	2
6	0	0	1	1	1
7	1	1	1	0	1
8	0	0	0	0	0
9	2	0	1	0	1
10	0	0	1	1	1
11	1	0	0	0	2
12	0	2	0	1	0
13	0	1	0	1	0
14	2	2	0	1	0
15	0	1	1	0	0
16	0	0	1	1	1
17	0	1	0	0	1
18	0	0	2	3	0
19	0	0	0	0	0
20	0	0	0	0	1
21	0	1	0	0	0
22	0	1	1	0	0
23	2	0	0	0	0
24	1	0	0	0	1
25	0	0	0	0	0

As mentioned in my approach, only the occurrence of one birthday pair was utilized in the data processing, as this is the aim of my investigation.

In the following table, I have determined the total occurrence of at least one birthday pair and the total occurrence of no birthday pairs, along with the corresponding percentages for each trial.

Table 4: Calculated Percentages for Single Pairs and No Pairs

Trial	Occurrence of Single Pairs	Occurrence of No Pairs	Percentage of Single Pairs	Percentage of No Pairs
1	9	16	36.0	64.0
2	10	15	40.0	60.0
3	10	15	40.0	60.0
4	7	18	28.0	72.0
5	11	14	44.0	56.0

Sample calculations for the percentage of single pairs in Trial 1 are included below.

$$\frac{9}{25} = 0.36$$

$$0.36 \times 100 = 36.0$$

Probability in percentage (sample)	36.0

Next, the average percentage was calculated for all four trials. Refer to methodology for more an explanation of why this is necessary. Sample calculations for the percentage of single pairs are included below.

$$\frac{36.0 + 40.0 + 40.0 + 28.0 + 44.0}{5} \times 100 = 37.6$$

Probability, in percentage, of one pair	37.6
Probability, in percentage, of no pairs	62.4

Thus, the investigation indicates that there is a 37.6% of one birthday pair given 23 people in a class. This information does not corroborate with the literature value of 50.0% as calculated and indicated in *Table 2*. However, this may be due to several reasons that are reflected below.

Small sample size – Although I attempted to use the entire high school population and conducted five trials, this may still not have been enough. This idea brings up the question of theoretical and experimental probability. Generally speaking, the more data collected, the more likely the probability will return to the literature value, or theoretical probability.

Use of technology – I am unable to completely verify the reliability of my online source in generating completely random set of data.

Omission of extra students from trials – This consideration led to increased variability amongst trials, which may have affected the probability of birthdays shared. This is a variable that was not controlled.

Processing for merely one birthday pair – Processing for only one birthday pair is part of the aim, but it ignores the fact that more than one birthday pair was apparent in a group, which surely would have increased the probability of shared birthdays.

Assumption of birthdays as independent – Although birthdays are typically independent of each other, in this particular case, they are dependent to a small degree because they remain within the parameters of my high school.

Surely, my approach to this investigation could have been attempted in a different manner. One example includes finding the number of birthday pairs when pairing all 607 students alone. This may attempt to alleviate the uncontrolled variables of omitting students from trials and processing for merely one birthday pair.

Conclusion

My original aim of this exploration included: calculating the probability of two people having the same birthday given n quantity of people and determining the probability that two people in my high school will share the same birthday.

My findings suggest that predetermined calculated values are much less apparent in reality. Probability and statistics make up a large portion of mathematics in the real-world, and thus, my investigation demonstrates how unlikely data can be distributed despite extensive probability calculations. Especially considering the approach employed to a given situation, results may deviate to a large degree.

To extend this investigation within the framework of the current aim, a comparison of different approaches, such as the one suggested above could expand on exploration. Beyond the current aim, this concept could be applied to determining the likelihood of somebody who comes from the same county as me, which may potentially be

The pigeonhole principle can be applied to many other examples in reality, such as computer sciences and solving algorithms. My application to the birthday problem merely represents a simple example of the implications that simple mathematics can have on complex issues.

Beyond the pigeonhole principle, the topic of probability in mathematics has far-ranging implications. Entire industries and economies have formed as a result of mathematical principles. After all, the gaming industry, lotteries, stocks, casinos, and casino cities would seize to exist without probability.

Bibliography

Azad, Kalid. "Easy Permutations and Combinations." *BetterExplained*. Better Explained, 2014. Web. 16 Sept. 2015.

Ben Dale, *The Pigeonhole Principle*. Wikipedia. July 2008, Web. 21 December 2015.

Combinations. By Salman Khan. Khan Academy, 2008. Online Video.

Dreamstime, *Chair Strokes Outline Vector Illustration*. Dreamstimes. 2014, Web. 20 September 2015.

Math and Multimedia, *Permutations (white and blue spaces) and Combinations (blue row)*. Math and Multimedia. 2010, Web. 20 December 2015.

Permutations. By Salman Khan. Khan Academy, 2008. Online Video.

Reese, George. "The Birthday Problem." *Office for Mathematics, Science, and Technology Education*. University of Illinois, 2015. Web. 21 Sept. 2015.

Simon, Scott. "Math Guy: The Birthday Problem." *NPR*. National Public Radio, 19 Mar. 2005. Web. 21 Sept. 2015.

Su, Francis. "Pigeonhole Principle." *Mudd Math Fun Facts*. Harvey Mudd College, 2010. Web. 18 Sept. 2015.

5. Identifying and comparing Dunkin Donut's

(assessment extracted from pdf)

Introduction

Torus is a circle of revolution or a three-dimensional figure which is formed by revolving a circle around an axis of revolution that doesn't intersect with it. The best example we could take of a figure like this is a doughnut, everyone around the world would know what this is since it's one of the most popular snacks ever. This is a great example as the doughnut has an iconic ring torus shape to it. Throughout history there have been a variety of doughnuts, some coating with chocolate, sugar, sprinkles, and various unimaginable combinations. But what makes a doughnut perfect? There are many famous companies like Krispy cream and Dunking doughnuts who are well known for the flavorsome doughnuts they make, but what actually makes the doughnut so good despite the toppings or ingredients used? What makes the doughnut have the right amount of crispiness without losing its soft pallet? This is the same question Dr. Eugenia Cheng wanted to know, therefore she conducted a study and derived the formula: $\frac{(r-2)^2}{4(r-1)}$. This looks at how the inner radius and the area of the doughnut hole causes the ratio of crispiness to softness change, she also derived the average surface area and volume the doughnut should have to complement the perfect value of 11mm obtained by her by using the above formula.

Aim

The aim of this exploration is to compare ring size (doughnut hole), surface area, and volume of a Dunking Doughnut with the values of the perfect doughnut obtained by Dr Eugenia Cheng.

I will be looking more into the torus figure in order to model the exact figure of an ideal doughnut. A general formula will be derived to calculate the ring size, surface area and volume according to the torus figure, these then will show the relation between ratio of crispiness to softness of a doughnut. Finally, I will compare the derived values of the Dunkin doughnut to the values of the perfect doughnut.

Rationale

When I first saw a Doughnut, its shape was very interesting to me, one of the reasons could be that it is a shape or figure I see often. Another reason is because it looks perfect and symmetrical, ever since I was a child I admired symmetrical objects and I wanted to know what makes such objects amusing or interesting to me. So, when I got the opportunity to answer this question with using mathematics I was all for it, the first thing I could think of was how I can integrate a doughnut with mathematics and found my answer when I looked at torus solid of revolution.

In order to achieve my aim, the first step if for me to model doughnuts from Dunkin Doughnuts to get the original readings for the shape, surface area and the volume of the doughnuts the store makes. I will be measuring six doughnuts to get the most accurate average values of the snack and also to model of the torus figure. Following this I will use the model of a doughnut and use calculus to get the area of the average doughnut whole, its surface area and finally its volume. After obtaining the values of the average doughnut I will compare it to the values of the perfect doughnut found out by Dr Eugenia Cheng. Furthermore, I will also look at how these values effect or may affect the crispiness to softness ratio.

Doughnut measurements and Identifying an equation

I took an average of six measurements to reduce errors, the results I've obtained are;

Height	Width
2.94	8.63
3.09	8.27
3.12	9.13
2.82	8.67
2.98	8.46
3.14	7.84

From this we calculate the average height and average width;

Average Height

$$\frac{2.94 + 3.09 + 3.12 + 2.82 + 2.98 + 3.14}{6} \approx 3.02$$

Average Width

$$\frac{8.63 + 8.27 + 9.13 + 8.67 + 8.46 + 7.84}{6} \approx 8.5$$

Now that we have calculated the average height and width of the doughnut, the next step is for the cross-section of torus figure to be plotted a graph, it will be symmetrical and identical circles so thearea would be two on opposite sides of the Y-axis. This image will be similar to that of Figure 1 below but will be in 2D.

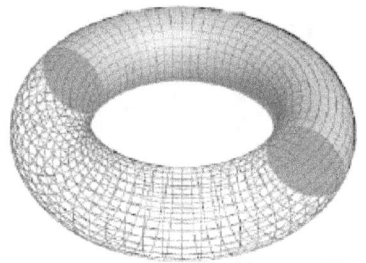
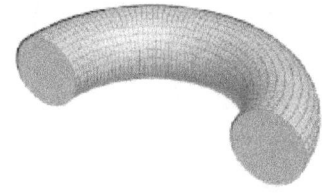

The general formula of the circle to be $(x-a)^2 + y^2 = z^2, z \in \mathbb{R}, z > 0$ where z represents the radius of the circle and a represents the distance from the circle to the y-axis. Hence to the cross section of the doughnut with the mentioned results of height and width above the equation of the circle must be translated and transformed in agreement to the radius of the cross section along with the distance between the y-axis to the cross section.

$(x - 2.679)^2 + y^2 = 1.328^2$

Due to the fact of the cross section having two sides we both the $+$ and $-$ because of the torus figure being a symmetrical and identical figure as stated above. Following this we will solve the equation to make y the subject of the equation in order to plot it on a graph.

$(x \pm 2.679)^2 + y^2 = 1.328^2$

$y^2 = 1.328^2 - (x \pm 2.679^2)$

$y = \pm\sqrt{1.328^2 - (x \pm 2.679)^2}$

As seen above, after y has been made the subject the equation can now be plotted on a graph where each unit represents one centimeter. Note that this equation denotes the exact cross section of a doughnut at a point where the radius (of the hole) is the biggest.

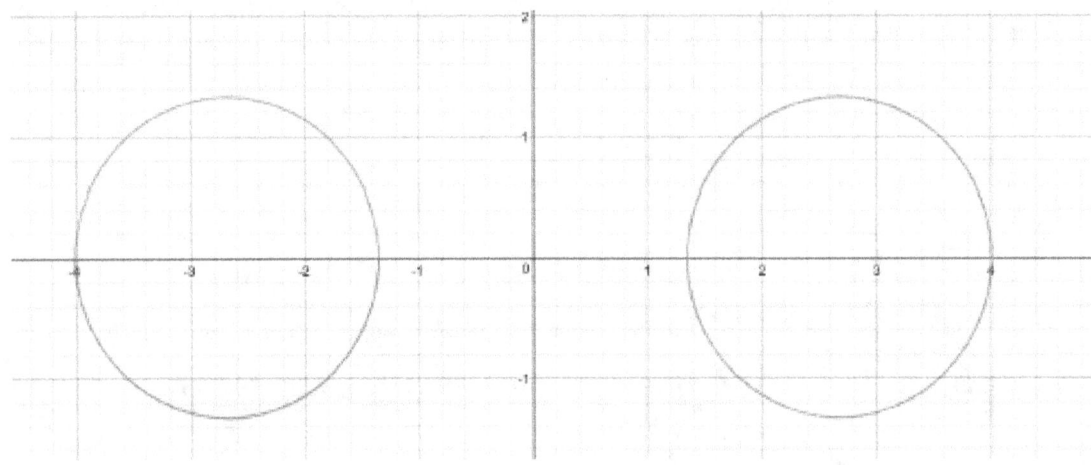

Figure 2 – Cross section of a doughnut (2d)

To further simplify calculations we can come up with general formulas to calculate the inner and outer radii of the torus figure. The values that can be calculated by using the general formulas will be crucial in this exploration and will be used in the upcoming steps.

$$(x \pm a)^2 + (y)^2 = z^2$$

$$a, z > 0 \quad a, z \in \mathbb{R}$$

$$a = \frac{R + r}{2} \quad \text{and} \quad z = \frac{R - r}{2}$$

$$Radius\ of\ the\ cross\ section = |z|$$

$$Outer\ radius = R = a + |z|$$

$$Inner\ radius = r = a + |z|$$

To identify the values of the inner and outer radius of the doughnut, we will use the above formulas.

$$Inner\ radius = 2.679 - |\sqrt{1.328^2}| = 1.35$$

$$Outer\ radius = 2.679 + |\sqrt{1.328^2}| = 4.01$$

$$Radius\ of\ the\ cross\ section = |\sqrt{1.328^2}| = 1.328$$

When modeling the equation for the cross section displayed in figure 2, there were many factors effecting the accuracy of the valued which lead to errors. One definite cause includes the values taken above are the averages and the measurements also may be inaccurate due to an absence in conformity of an actual doughnut sold in Dunkin Doughnuts. In addition, the doughnut will have to be cut perfectly to obtain the cross section as expected in the graph above.

Modeling doughnut with measurements taken

The first step to take while modelling the doughnut is to revolve the circle with the dimensions close to its cross-section throughout the axis of revolution. To do this the equation of the cross-section from earlier will be used, but only the positive side or rather the right side will be taken in account because when revolving we only look at one side of the axis of revolution as having two sides make it too complex.

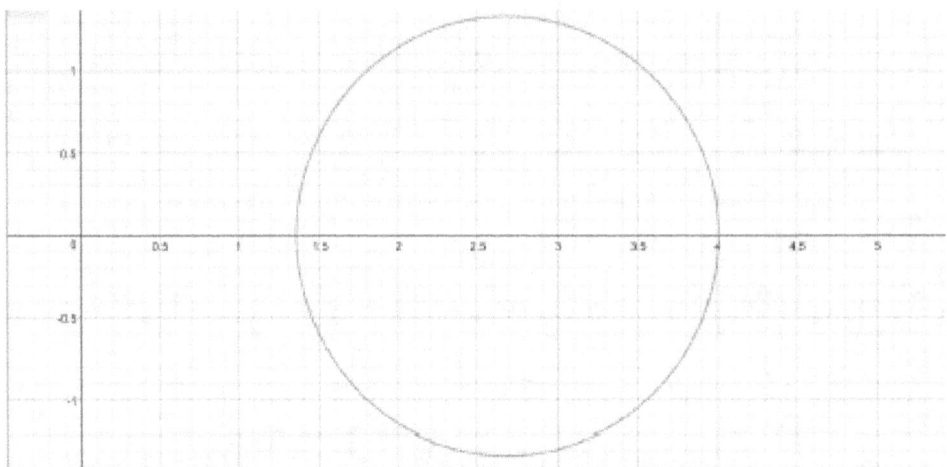

The above equation will be rotated around the y-axis in order to model the doughnut with the measurements obtained. The equation for the doughnut will be converted into a 3-dimensional Cartesian figure with the formula:

$$(a - \sqrt{x^2 - y^2})^2 + c^2 = z^2$$

In this formula a signifies the distance between the center of the hole of the torus figure to the cross-section of the circle. The radius of the cross-section in this case is represented by z. Now we substitute the values given above in the equation to get a torus figure such like the one given below.

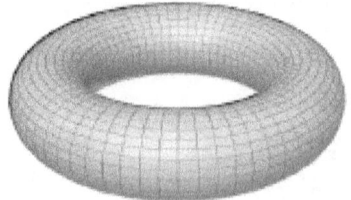 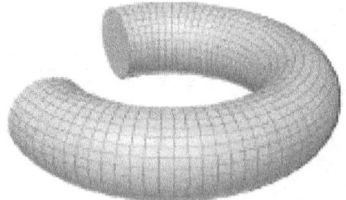

Deriving formula for the surface area of a torus figure

The equations given below will be used to calculate and determine the surface area of a torus, because this is the equation of the circle that will be rotated around the axis to form or rather shape the torus. (Blaavand, 2006)

$$f(y) = a \pm \sqrt{c^2 - y^2} \quad f'(y) = \pm \frac{y}{\sqrt{c^2 - y^2}}$$

The equation above will then be used in the formula below to calculate the surface area of the torus after the circle has revolved about the y-axis. (Blaavand, 2006)

$$Surface\ of\ revolution = 2\pi \int_a^b f(y)\sqrt{1 + f'(y)^2}\, dx$$

As the doughnut is symmetrical, we just need to use the above formula and revolve the upper half of the torus figure multiplying it by two.

$$Area = 4\pi \int_{-b}^{b} (a + \sqrt{z^2 - y^2})\sqrt{1 - \frac{y^2}{z^2 - y^2}}\, dx$$

$$Area = 4\pi \int_{-b}^{b} a\sqrt{\frac{z^2 - 2y^2}{z^2 - y^2}} + \sqrt{z^2 - 2y^2}$$

$$Area = 4\pi ac \left[\arctan\left(\frac{y}{\sqrt{z^2 - y^2}}\right)\right]_{-b}^{b} = 4\pi^2 ac$$

$$Hence,\ Area = \pi(R + r)(R - r)$$

Since we already know the values of the inner and outer radius we can use this general formula to find the approximate surface area of the average Dunkin Doughnut.

$$Surface\ area = \pi(4.01 + 1.35)(4.01 - 1.35)$$

$$\approx 22.2\ cm^2$$

When using this method, we are assuming that the doughnut is having a continuous surface area and that it has an even surface throughout. However, this is most likely not the case as companies can't really produce a doughnut completely even with a continuous surface area, but this formula gives us a general idea.

Deriving the volume of the doughnut

The first step in order to derive the formula for the volume of the torus figure is to make x the subject. (Blaavand, 2006)

$$(x-a)^2 + y^2 = z^2$$
$$(x-a)^2 = z^2 - y^2$$
$$x - a = \pm\sqrt{z^2 - y^2}$$
$$f(y) = x = a \pm \sqrt{z^2 - y^2}$$

Now we use the volume of revolution:

$$\pi \int_a^b f(y)^2 dy$$

We can now derive the formula for the volume we can just revolve the upper half of the circle about the y-axis in order to get the general formula where we can then input the values of the inner and outer radius for the calculations.

$$Volume = \pi \int_{-d}^{d} (a + (\sqrt{c^2 - y^2})^2 - (a - (\sqrt{c^2 - y^2}))^2 \, dy$$

$$= \pi \int_{-d}^{d} a^2 + c^2 \, dy + 2a\pi \int_{-d}^{d} \sqrt{c^2 - y^2} \, dy - \pi \int_{-d}^{d} y^2 \, dy - \pi \int_{-d}^{d} a^2 + c^2 \, dy$$

$$+ 2a\pi \int_{-d}^{d} \sqrt{c^2 - y^2} \, dy + \pi \int_{-d}^{d} y^2 \, dy$$

$$Hence, = 4a\pi \int_{-d}^{d} \sqrt{c^2 - y^2} \, dy \quad taking\ (1)\ y = c(sinx) (2) \frac{dy}{dx} = c(cosx)$$

$$= 4a\pi \int_{-\frac{\pi}{2}}^{\frac{\pi}{2}} \sqrt{c^2 - c^2(sinx)^2} \, dy = 4a\pi \int_{-\frac{\pi}{2}}^{\frac{\pi}{2}} c(cosx) dy \quad when\ substituting\ 1$$

$$by\ using\ cos^2 x = 2cos2x - 1\ (double\ angle\ formula)$$

$$= 4a\pi \int_{-\frac{\pi}{2}}^{\frac{\pi}{2}} c^2(cosx)^2 dx = 4ac^2\pi \int_{-\frac{\pi}{2}}^{\frac{\pi}{2}} \frac{cos2x + 1}{2}$$

$$= 4ac^2\pi \int_{-\frac{\pi}{2}}^{\frac{\pi}{2}} \left[\frac{sin2x}{4} + \frac{x}{2}\right] = 2ac^2\pi \left[\frac{sin(\pi)}{2} + \frac{\pi}{2} - \frac{sin(-\pi)}{2} + \frac{\pi}{2}\right]$$

$$2ac^2\pi^2 = 2\pi^2 (\frac{R+r}{2})(\frac{R-r}{2})^2$$

$$Volume = \frac{\pi^2}{4}(R+r)(R-r)^2$$

Just like with the surface area we can now get the approximate volume of the Dunkin Doughnut's doughnut by using the values of R and r from the ones calculated above.

$$Volume = \frac{1}{4}\pi^2(4.01 + 1.35)(4.01 - 1.35)^2$$
$$\approx 93.6 cm^3$$

The volume obtained above by using the formula in this case is measured by using the cross-section of one part of the doughnut/torus figure and so the sharpness or rather accuracy can be effected due to possible errors made when modeling the doughnut. In addition, when integrating, the volume does not consider the small circle openings that are formed when the doughnut is baked. This shouldn't be a big problem as most people do not consider the circular openings when measuring the volume of a doughnut.

Deriving a formula to calculate the area of a circle using calculus

To start we need the basic equation of the circle which is $x^2 + y^2 = a^2$, when making y as the subject we get $y = \pm\sqrt{a^2 - x^2}$. (Analyzemath)

Knowing this the equation of the upper semi-circle is given by

$$y = \sqrt{a^2 - x^2}$$

$$= a\sqrt{\frac{1-x^2}{a^2}}$$

By adding integrals, we can know find the area of the upper right quarter of the circle as seen below

$$\frac{1}{4} Area\ of\ circle = \int_0^a a\sqrt{\frac{1-x^2}{a^2}} dx$$

Now if we substitute the above formula with sin and cos we get:

$$\frac{1}{4} Area\ of\ circle = \int_0^{\frac{\pi}{2}} a^2(\sqrt{1-sin^2 t})\cos t\, dt$$

$$= \left(\sqrt{1-sin^2 t}\right) = \cos t$$

$$\frac{1}{4} Area\ of\ circle = \int_0^{\frac{\pi}{2}} a^2 \cos^2 t\, dt$$

$$\frac{1}{4} Area\ of\ circle = \int_0^{\frac{\pi}{2}} a^2 \frac{\cos 2t + 1}{2} dt$$

$$\frac{1}{4} Area\ of\ circle = \frac{1}{2}a^2 \int_0^{\frac{\pi}{2}} \frac{1}{2}\sin 2t + t$$

$$\frac{1}{4} Area\ of\ circle = \frac{1}{2}a^2 \left[\left(\frac{1}{2}\sin 2\left(\frac{\pi}{2}\right) + \frac{\pi}{2}\right) - \left(\frac{1}{2}\sin 2(0) + 0\right)\right]$$

$$= \frac{1}{4}\pi a^2$$

Since the total area of the circle is obtained by multiplying 4 to the equation above so the $\frac{1}{4}$ is cancelled. Hence the area of the circle is;

$$Area\ of\ circle = \pi a^2$$

Similar to what we did with the surface area we just need to input the inner radius value into the equation to find the area of the doughnut hole which looks like

$$Area\ of\ circle = \pi(1.35^2)$$

$$= 5.72mm$$

While looking at the area of the doughnut hole we assume its symmetrical and perfect but this is not the case, in an actual doughnut the doughnut hole may not be shaped in a pure circular form as there could be some problems due to doughnuts being produced in batches. In addition, similar to the volume and surface area there could be additional errors in the area of the doughnut hole from the value of a because of previously made errors when modelling the doughnut.

How Surface area Volume and Area of doughnut hole effect the ratio of Crispiness to Softness

The first aspect and the biggest that effects the ratio is the area of the doughnut hole, more specifically the radius. According to Dr Eugenia Cheng and the given formula below the perfect doughnut should have a radius of 11mm or 0.4 inches.

$$\frac{(r-2)^2}{4(r-1)}$$

The radius or area of the doughnut hole plays a more significant role in the ratio compared to the surface area as it directly impacts the softness or crispness as seen in the above formula. If the radius is too small the doughnut will be soft and if the radius is too big the doughnut will be too crispy. Looking at our findings from the Dunkin doughnut we can see that the radius and area of the doughnut hole are considerably smaller than the perfect size which is no surprise as there are various factors effecting the size and shape of the doughnut such as baking time, type of dough used, toppings, etc. Basically, there is no real way of making the perfect doughnut but Dunkin Doughnuts can defiantly use this information to come close.

Coming to the surface area, it does not have a direct impact on the crispiness to softness ratio for the doughnut but it has an impact on the radius which the ratio depends on, if the surface area is too big then the radius tends to be bigger the radius is smaller if the surface area is smaller. So, a higher surface area will lead to a crisper doughnut whereas a smaller surface area doughnut will make it softer, there is no real number for what the surface has to be as different companies make a variety of sized doughnuts but a number around $35cm^2$ is good. Hence by looking at our findings we can presume that this exploration will help Dunkin Doughnuts improve their quality of the goods they provide as well as help them get a better idea of how a perfect doughnut should look like or rather the dimensions it should consist of to give the consumer the best experience.

Finally, the volume similar to the surface area doesn't play a big part in the crispiness to softness ratio but effects it none the less. If the doughnut has a very high volume it means that it leans to the more softer side whereas if the volume is less the doughnut becomes crispier. The ideal and acceptable volume for a doughnut is between $75cm^3$ to $86cm^3$ and anything higher or lower than that could cause an unfavorable ratio between the crispiness and softness of the doughnut.

Reflection

Looking at the method used throughout this exploration, we have seen several drawbacks and benefits. When we modelled the doughnut's surface area and volume we assumed that the doughnut itself was a perfect torus figure to avoid any complicated errors. This most probably have had led to various blunders when we moved to the next stage to calculate the cross-section as it may have varied when taken at different points in the figure. In addition, when the doughnut is put through the baking process using dough, it increases in size or rather expands from its previous shape in an uncertain randomized nature hence preventing the surface of the figure from being curved exactly like a torus. Even when we took account of all these uncertainties, there still could be a possibility that the measurements that were taken are inaccurate as they could be a little off. But, in order to minimize this, we took an average value in order to reduce the uncertainties of human error or parallax error in this case.

Coming to the real-life problems, big companies like Dunkin doughnuts may not be able to change their production process as most of these big companies mass produce daily which makes it very costly to alter it.

We can improve the accuracy of this exploration by measuring more doughnuts to get an accurate average value or by considering how the volume of toppings such as sugar or chocolate effect the torus figure. By adding this we could consider how the toppings effect the ratio between crispiness to softness and also look into how the volume changes when using various different doughs to get a more accurate and clearer picture of the process.

To conclude, looking at the scope of this exploration, we can say that results gained have a great deal of implications when we look at how Dunkin doughnuts can improve the quality of their goods to beat out their competitors. This report or research can be used as a base for any further more complex research into the same or similar topic to get better and more accurate results due to using various other methods gathered from different people. In this explorations case, it is limited by various amounts of factors which could have led to errors for Dunkin doughnuts to make the closest possible match to the perfect doughnut and so caution must be taken if these explorations findings are to be used as it needs some room for space. In addition, this exploration does not include the derivation of the perfect doughnut formula as there is very limited information online from Doctor Cheng, this may hinder the understanding of what the exact factors effecting this formula are. Hence finding out this information may further develop the potency of this research paper and will improve the overall findings of this exploration.

Work Cited

Blaavand, Jakob Lindblad. Article title "Surface area and volume of a torus.", Website title "Surface area and volume of a torus."Aarhus University, Date Accessed February 20th 2017. URL http://home.math.au.dk/blaavand/torus.pdf

No Author, Article Title **Find The Area of Circle Using Integrals in Calculus**, Website Title **Find The Area of Circle Using Integrals in Calculus**, Date Accessed **February 20, 2017**, URL http://www.analyzemath.com/calculus/Integrals/area_circle.html

Tom Mulvihill ,Article Title **Mathematician creates formula for the perfect doughnut**, Website Title **The Telegraph**, Date Published December 10, 2014, Date Accessed **February 20, 2017**, URL http://www.telegraph.co.uk/foodanddrink/foodanddrinknews/11284809/Mathematician-creates-formula-for-the-perfect-doughnut.html

No Author, Article Title **Torus**, Website Title **Torus -- from Wolfram Math World**, Date Accessed **February 20, 2017**, URL http://mathworld.wolfram.com/Torus.html

No Author, Website Title **study.com**, Date Accessed **February 20, 2017**, URL http://study.com/academy/lesson/volume-surface-area-of-a-torus.html

6. Ciphers and their Solutions

Structure of the investigation

The aim of this investigation was to research and communicate an understanding of the mathematical principles behind Banburismus, a method used by British cryptanalysts in the Second World War to decrypt Enigma codes. I began by using probability in the form of the binomial coefficient to calculate the total number of possible Enigma settings. I then investigated the Caesar and Vigenère ciphers to learn about frequency analysis of letters as a method of cipher solving. I looked at different ways of solving the Vigenère cipher, from modular arithmetic (which is only successful if the key word is known) to the Kasiski Examination and the Friedman Test. This led me to research Friedman's index of coincidence, which was useful in introducing me to the idea of matching characters in two sets of text as indication of their having been written with the same alphabet. I finished by investigating Bayes' Theorem of Probability. This allowed me to combine my research into basic codes and their mathematics with a more complicated equation that was actually used to crack Banburismus.

Personal engagement with the topic

As part of my IB Diploma, I study History SL and German HL as well as Maths SL. In History, I found learning about the Second World War very interesting, but was eager to extend my knowledge beyond the course focus. Having watched 'The Imitation Game' and learnt a little about the role of Bletchley Park in the War, I was keen to link my study of Maths and History together by investigating the mathematics behind ciphers and their solutions. In particular, I wanted to apply my interest in probability (an area of maths which I knew had many practical uses) to a new challenge. This also tied in with my interest in languages, as there are many similarities between the intricacies of grammar and linguistics and those of cryptanalysis, and I found it interesting how one letter can symbolise another by a complex procedure of computing. Equally, the Enigma machine was a German invention and its messages were transmitted in German, which interested me because I wanted to further my understanding of German language and culture.

The context of Banburismus

Alan Turing was a British mathematician born in 1912, who took up a job at Bletchley Park in 1939, where confidential work was being carried out to crack German war codes (Clements, 2016). Whilst at Bletchley, Turing focused on cracking the 'Enigma' code – the Enigma was a type of enciphering machine used by the Germans. The Poles had already cracked the Enigma, but the Germans increased their security measures at the onset of war by changing the cipher system daily. Hence, Poland asked Britain for help (Clements, 2016).

Turing, along with fellow mathematician and code-breaker Gordon Welchmann, invented a machine known as the Bombe, which significantly reduced the work that the code-breakers needed to put in (Clements, 2016). From mid 1940, German air signals were being successfully read at Bletchley Park, which greatly helped the war effort.

Turing also worked on trying to decode the more complicated German naval signals, as German submarines were inflicting heavy losses on allied ships. These naval messages could be read from 1941 after Turing and his 'Hut 8' team developed a cryptanalytic technique called Banburismus that helped greatly during the Battle of the Atlantic by directing allied convoys away from the U-boat 'wolf-packs'. Banburismus reduced the time required to operate the Bombe machine by identifying the most likely right-hand and middle wheels of the Enigma (Clements, 2016).

The functioning of the Enigma machine

The Enigma Machine looked like a typewriter keyboard; the German intelligence would input their message via the standard keys (Hern, 2014).

The signal first passes through three rotors (Hern, 2014), which change the output letter, and then a reflector sends the signal back through the whole system, to link the original letter with the new coded letter. The first rotor rotates one step after each key press, and after 26 presses, the second rotor starts to move (Hern, 2014). The signal would then pass through a plugboard that switches the letter for a different one, adding an extra layer of complexity to the code (Bryon, 2015). The final encrypted letter would light up on the lampboard.

However, the operators only needed the information about the starting position and order of the three rotors, as well as the position of the plugs in the board, to decode the text.

Photograph of the Enigma machine

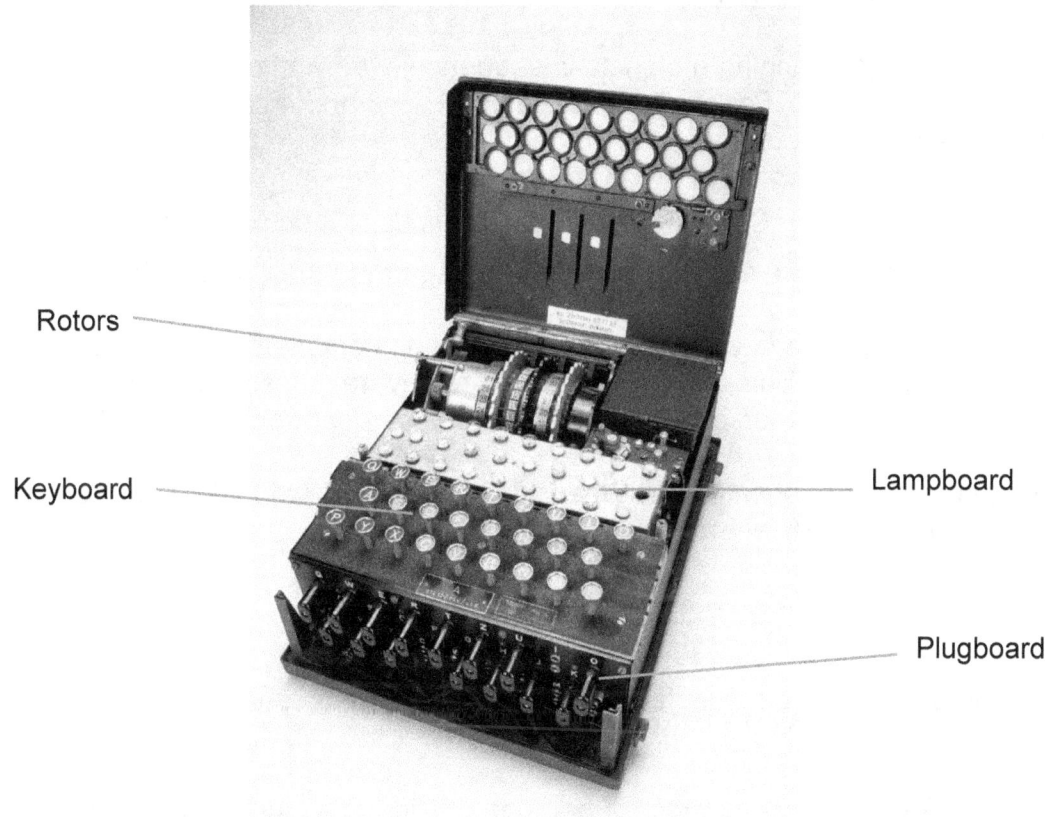

Figure 2

Using Banburismus to break Enigma

The reflector component made the Enigma machine reciprocal, so if A was enciphered to B on a specific setting, then B would be enciphered to A (Hern, 2014). This meant that the machine did not need to be constantly changed from 'encipher' to 'decipher' mode. However, once the Allies discovered that no letter could ever be encoded as itself (Hern, 2014), the Enigma code became considerably easier to solve.

Banburismus used sequential conditional probability to work out the likely settings of the Enigma machine (Tarran, 2014). This gave rise to Turing's invention of the 'ban' as a measure of the weight of evidence in favour of a hypothesis (Wikipedia).

Counting the possible plugboard settings

The basic Enigma machine has three rotors, each of which has 26 possible positions (Sale, n.d.).

26 x 26 x 26 = 17576
3 ! = 6

As 3 ! is equal to 6, there are six different ways of arranging three objects, and therefore there are six different ways that the three rotors can be positioned in.

17576 x 6 = 105456

So, altogether the Enigma machine can be positioned in 105456 different ways.

Next, we need to take into account the plugboard settings, which add an extra layer of complexity to the Enigma code.

There are n! ways of arranging n objects in sequence, where n! means n factorial, or n x (n – 1) x (n – 2) x (n – 3) etc. For example, the digits 1, 2, 3 and 4 can be arranged in 4 x 3 x 2 x 1, or 24, different orders.

This leads to the formula: $^nC_r = \frac{n!}{r!(n-r)!}$, which allows us to calculate the number of ways of choosing r elements from a set of n elements, disregarding their order (Weisstein, n.d.).

On the Enigma plugboard, there are 10 connected pairs of letters and 6 letters that are unpaired (Hodges, u.d.). This means that there are several possible settings of the machine that need to be tried before a message can be decoded.

To work out the number of settings on the Enigma machine, we first have to choose 2 letters, or a pair, out of 26 letters.

$$^{26}C_2 = \frac{26!}{2!(26-2)!} = \frac{26!}{2!(24)!} = \frac{26 \times 25}{2 \times 1} = 325$$

The next two options will be selected out of the remaining 24 letters.
$^{24}C_2 = 276$

In order to work out the total number of pair possibilities, we multiply all of these values together (Hodges, u.d.), until there are 6 individual letters remaining.

$^{26}C_2 \times {}^{24}C_2 \times {}^{22}C_2 \times \ldots {}^{8}C_2$

$$= \frac{26 \times 25}{2} \times \frac{24 \times 23}{2} \times \frac{22 \times 21}{2} \times \frac{20 \times 19}{2} \times \frac{18 \times 17}{2} \times \frac{16 \times 15}{2} \times \frac{14 \times 13}{2} \times \frac{12 \times 11}{2} \times \frac{10 \times 9}{2} \times \frac{8 \times 7}{2}$$

$$= \frac{26!}{6! \, 2^{10}}$$

These 10 pairs can go in many different orders (Hodges, n.d.), and this number is denoted by 10!, or 10 factorial. In order to account for this, we need to divide the current formula by 10! (Hodges, n.d.).

$$\frac{26!}{6! \, 10! \, 2^{10}} = \frac{4.032914611 \times 10^{26}}{720 \times 3628800 \times 1024} = 1.507382749 \times 10^{14}$$

This means that there are 150,738,274,900,000 plugboard settings.

We can also use this information to create a general formula for the choosing of m pairs out of n objects.

C (n, m) = $\frac{n!}{(n-2m)!\, m!\, 2^m}$ (Hodges, u.d.)

For example, if we want to find the number of ways of choosing one pair of letters out of fourteen letters:

C (14, 2) = $\frac{14!}{(14-4)!(2)!(2)^2}$ = $\frac{14 \times 13 \times 12 \times 11}{2 \times 1 \times 4}$ = $\frac{24024}{8}$ = 3003

The Caesar cipher

I had hoped to simulate an Enigma code by understanding the trigram system and therefore using the indicators. However, upon researching the complex settings and components of the Enigma machine, and calculating the massive number of possible settings that exist, I realised that this would probably be beyond my comprehension. The Enigma machine is extremely complicated and Turing and his team employed PhD-level mathematics in creating Banburismus. I therefore decided to take some of the elements of Enigma to create some simple codes, and work out different ways to crack them.

The Enigma code is a complex mixture of more simple substitution ciphers (Sale, n.d.). A Caesar Cipher involves replacing each letter of the message to be encrypted with a different letter that is a fixed number of positions away in the alphabet (Ciphermachines.com, (n.d.)).

A B C D E F G H I J K L M N O P Q R S T U V W X Y Z
    A B C D E F G H I J K L M N O P Q R S T U V W X Y Z

So E would encrypt to A, F would encrypt to B, and so on.

This can be more easily represented as a wheel, because the fixed number of positions between the alignments of the two alphabets can be changed.

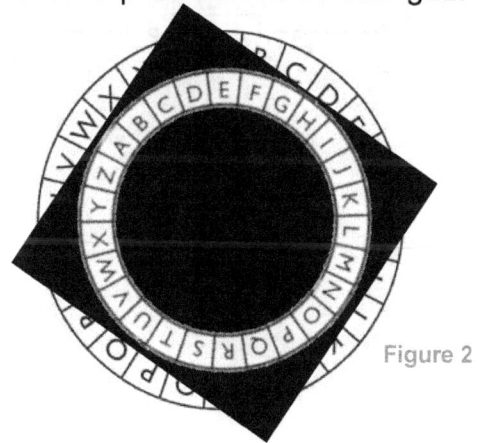

Figure 2

I am going to encrypt a simple German sentence with meaning: Yesterday I went to Berlin. As German naval messages would have naturally been encrypted in German, I wanted to extend my enjoyment of foreign language learning into the context of coding. I split the phrase into groups of 5 capital letters because this is how Enigma messages were transmitted to remove the semblance of words of different length (Rijmenants, 2004).

GESTE RNBIN ICHNA CHBER LINGE FAHRE N
CAOPA NJXEJ EYDHW YDXAN HEJCA BWDNA J

To solve this cipher, you need to look at the frequency distribution of letters in the coded phrase. I wanted to investigate the frequency distribution of letters in German. I chose three short random passages from German books, *Der Vorleser*, *Der Prozess* and *Deutschstunde*, and made a count of how many times each letter appeared in each extract. I took the German character 'ü' to be a standard 'u' and 'ß' to mean 'ss' as this is how they would have been encrypted for ease of use of the Enigma machine.

'Der Vorleser' (Appendix A)

A	B	C	D	E	F	G	H	I	J	K	L	M	N	O	P	Q	R	S	T	U	V	W	X	Y	Z
22	4	7	11	31	12	4	19	17	0	1	5	9	15	5	1	0	21	15	20	11	4	2	0	0	2

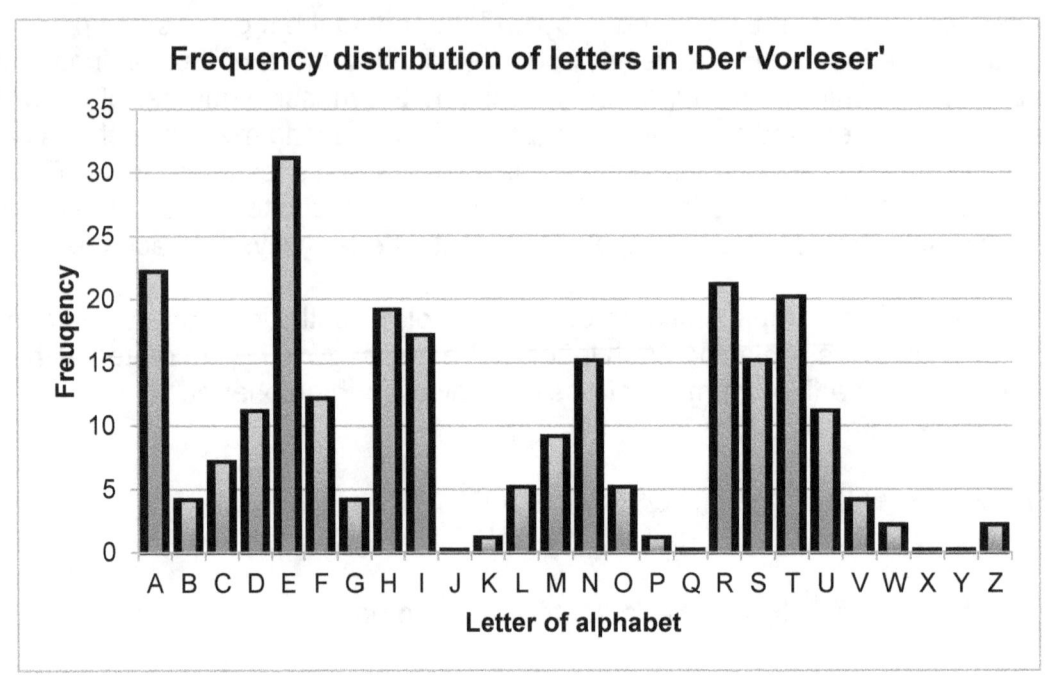

Figure 3

'Der Prozess' (Appendix B)

A	B	C	D	E	F	G	H	I	J	K	L	M	N	O	P	Q	R	S	T	U	V	W	X	Y	Z
18	4	8	12	40	5	7	16	12	3	4	3	11	17	6	0	0	18	17	14	9	3	3	0	0	1

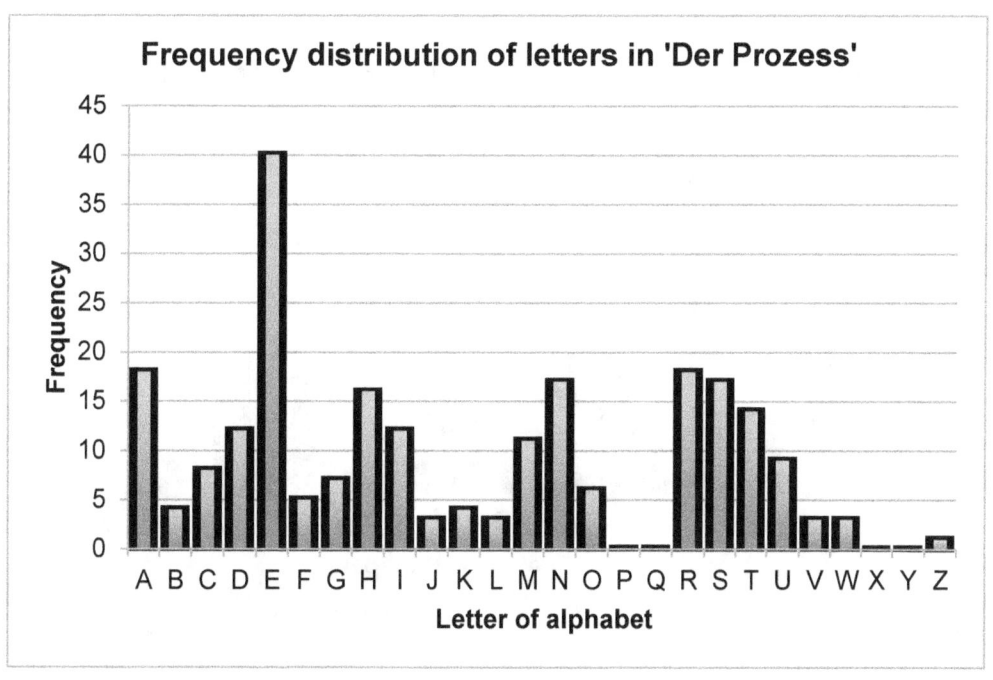

Figure 4

'Deutschstunde' (Appendix C)

A	B	C	D	E	F	G	H	I	J	K	L	M	N	O	P	Q	R	S	T	U	V	W	X	Y	Z
12	6	11	3	37	2	9	16	29	0	6	10	8	21	4	2	0	13	18	11	13	0	5	0	0	4

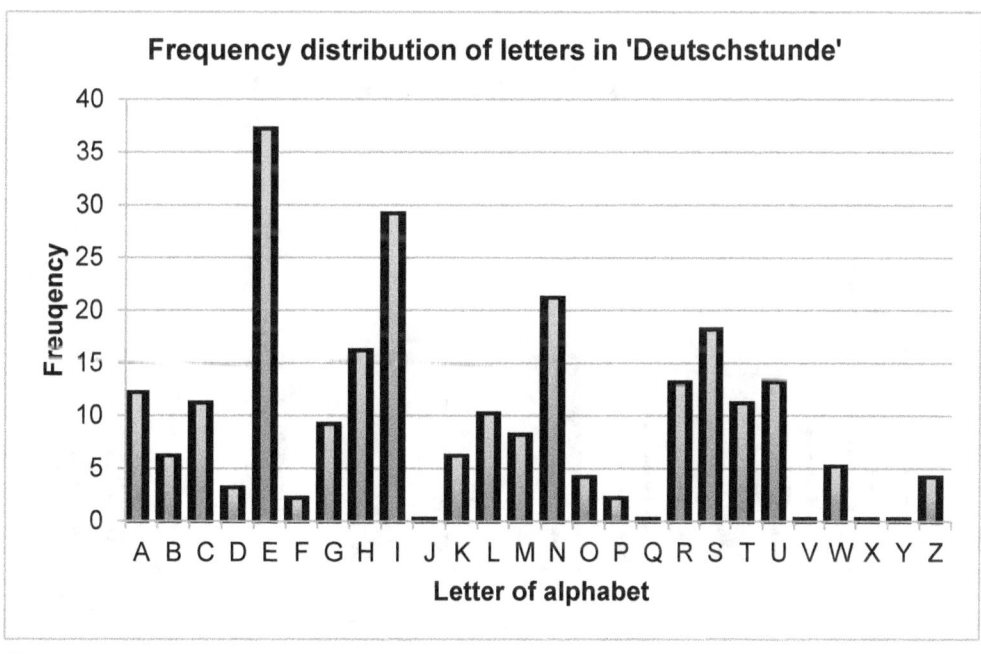

Figure 5

I counted up the total number of each letter in these three passages, out of 711 characters overall, and used the analysis to assume information about the frequency distribution of letters in the German language as a whole.

A	B	C	D	E	F	G	H	I	J	K	L	M	N	O	P	Q	R	S	T	U	V	W	X	Y	Z
53	14	26	26	108	19	20	51	58	3	11	18	28	53	15	3	0	52	52	45	32	7	10	0	0	7

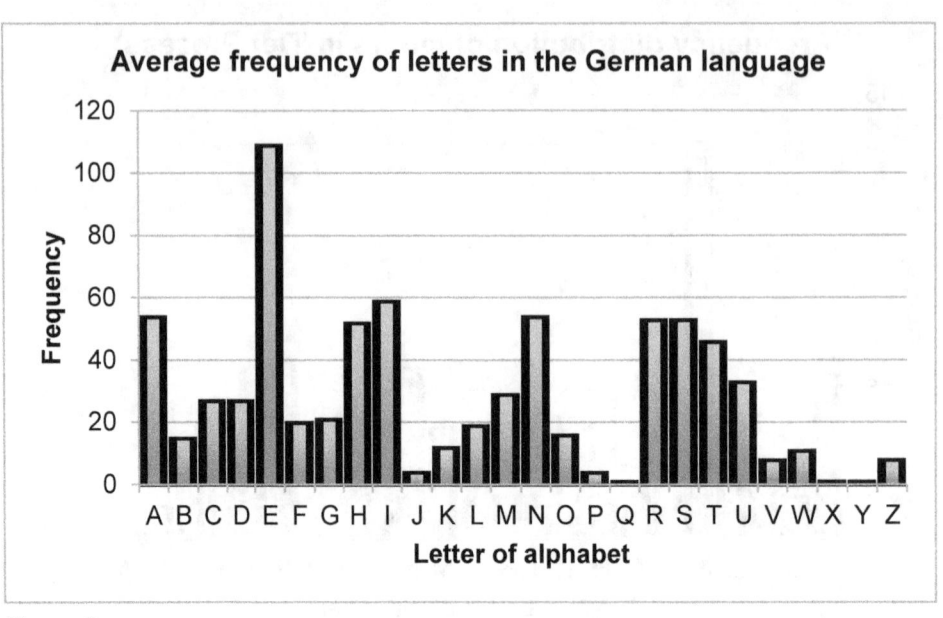

Figure 6

The results show that 'E' is the most common letter in the German alphabet by a long way.
If we return to the encrypted phrase: CAOPA NJXEJ EYDHW YDXAN HEJCA BWDNA
and make a similar count of the frequency of the letters, we get:

A	B	C	D	E	F	G	H	I	J	K	L	M	N	O	P	Q	R	S	T	U	V	W	X	Y	Z
5	1	2	3	3	0	0	2	0	3	0	0	0	3	1	1	0	0	0	0	0	0	2	2	2	0

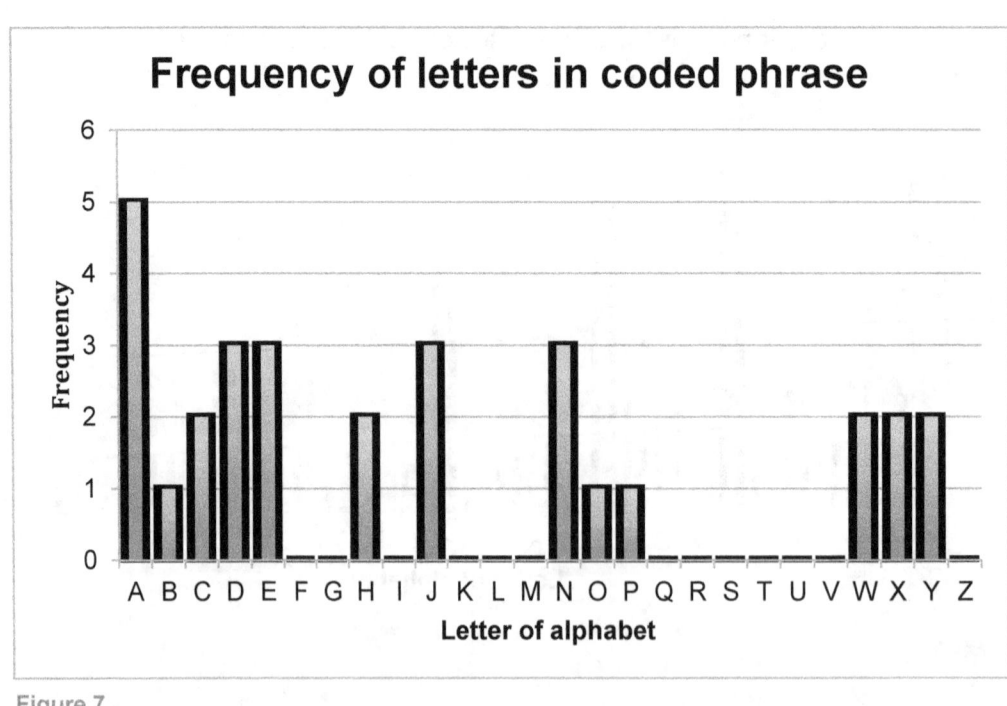

Figure 7

The letter 'A' has the highest frequency, so we could reasonably assume that 'A' in the cipher alphabet represents 'E' in the normal alphabet.

The other letters with high frequencies in the normal German alphabet are, from my calculations: $I = 58$, $N = 53$, $A = 53$, $R = 52$, $S = 52$, $H = 51$

In the encrypted alphabet, D, E, J and N all have a frequency of 3. It seems reasonable to check if 'H' and 'I' encrypt to 'D' and 'E' because 'H' and 'I' are consecutive letters in the alphabet which both have relatively high frequencies. If we also assume that 'N' has been encrypted to 'J', then the code looks like this:

CEOPENNXINIYHHWYHXENHINCEBWHNEN
_E_ _E_ _ N _ INI_H_ _ _H _ E _ _ IN _ E _ _ H _ E N

The Caesar cipher is the easiest to use, so it would be the first method that cryptanalysts would check for. From this information, they would definitely be able to crack the code.

Polyalphabetic ciphers

I wanted to extend the complexity of my research into ciphers. Polyalphabetic ciphers change the Caesar Cipher within a message, which makes it much more difficult to break a code using frequency analysis alone. This can be represented by the Vigenère Cipher Disk (Wikipedia, 2016), which consists of a stationary outer wheel and a rotating inner wheel, or by using a table like the one below:

	A	B	C	D	E	F	G	H	I	J	K	L	M	N	O	P	Q	R	S	T	U	V	W	X	Y	Z
A	A	B	C	D	E	F	G	H	I	J	K	L	M	N	O	P	Q	R	S	T	U	V	W	X	Y	Z
B	B	C	D	E	F	G	H	I	J	K	L	M	N	O	P	Q	R	S	T	U	V	W	X	Y	Z	A
C	C	D	E	F	G	H	I	J	K	L	M	N	O	P	Q	R	S	T	U	V	W	X	Y	Z	A	B
D	D	E	F	G	H	I	J	K	L	M	N	O	P	Q	R	S	T	U	V	W	X	Y	Z	A	B	C
E	E	F	G	H	I	J	K	L	M	N	O	P	Q	R	S	T	U	V	W	X	Y	Z	A	B	C	D
F	F	G	H	I	J	K	L	M	N	O	P	Q	R	S	T	U	V	W	X	Y	Z	A	B	C	D	E
G	G	H	I	J	K	L	M	N	O	P	Q	R	S	T	U	V	W	X	Y	Z	A	B	C	D	E	F
H	H	I	J	K	L	M	N	O	P	Q	R	S	T	U	V	W	X	Y	Z	A	B	C	D	E	F	G
I	I	J	K	L	M	N	O	P	Q	R	S	T	U	V	W	X	Y	Z	A	B	C	D	E	F	G	H
J	J	K	L	M	N	O	P	Q	R	S	T	U	V	W	X	Y	Z	A	B	C	D	E	F	G	H	I
K	K	L	M	N	O	P	Q	R	S	T	U	V	W	X	Y	Z	A	B	C	D	E	F	G	H	I	J
L	L	M	N	O	P	Q	R	S	T	U	V	W	X	Y	Z	A	B	C	D	E	F	G	H	I	J	K
M	M	N	O	P	Q	R	S	T	U	V	W	X	Y	Z	A	B	C	D	E	F	G	H	I	J	K	L
N	N	O	P	Q	R	S	T	U	V	W	X	Y	Z	A	B	C	D	E	F	G	H	I	J	K	L	M
O	O	P	Q	R	S	T	U	V	W	X	Y	Z	A	B	C	D	E	F	G	H	I	J	K	L	M	N
P	P	Q	R	S	T	U	V	W	X	Y	Z	A	B	C	D	E	F	G	H	I	J	K	L	M	N	O
Q	Q	R	S	T	U	V	W	X	Y	Z	A	B	C	D	E	F	G	H	I	J	K	L	M	N	O	P
R	R	S	T	U	V	W	X	Y	Z	A	B	C	D	E	F	G	H	I	J	K	L	M	N	O	P	Q
S	S	T	U	V	W	X	Y	Z	A	B	C	D	E	F	G	H	I	J	K	L	M	N	O	P	Q	R
T	T	U	V	W	X	Y	Z	A	B	C	D	E	F	G	H	I	J	K	L	M	N	O	P	Q	R	S
U	U	V	W	X	Y	Z	A	B	C	D	E	F	G	H	I	J	K	L	M	N	O	P	Q	R	S	T
V	V	W	X	Y	Z	A	B	C	D	E	F	G	H	I	J	K	L	M	N	O	P	Q	R	S	T	U
W	W	X	Y	Z	A	B	C	D	E	F	G	H	I	J	K	L	M	N	O	P	Q	R	S	T	U	V
X	X	Y	Z	A	B	C	D	E	F	G	H	I	J	K	L	M	N	O	P	Q	R	S	T	U	V	W
Y	Y	Z	A	B	C	D	E	F	G	H	I	J	K	L	M	N	O	P	Q	R	S	T	U	V	W	X
Z	Z	A	B	C	D	E	F	G	H	I	J	K	L	M	N	O	P	Q	R	S	T	U	V	W	X	Y

Figure 8

The coder chooses a key word (Wikipedia, 2016) – I will use *Birne*, German for 'pear'.

The message I will encrypt is the same as before, but with two additional sentences: *Gestern bin ich nach Berlin gefahren. Ich habe den Markt besucht und ich habe auch viele Sehenswürdigkeiten gesehen. Es hat viel Spaß gemacht und ich hoffe, dass ich bald wieder zurückkehren kann.*

The key word needs to be repeated until it replicates the number of letters in the message to be encoded, so this will be (*BIRNE* x 36) + *BI*.

Each row of the Vigenère Square starts with a key letter (Wikipedia, 2016), and each letter of the message will be enciphered using its corresponding key row, so only five rows of the Square are actually used with my key word. I have used the letter of the key word as the row reference, and the letter to be encrypted as the column reference. The row and the column intersect on the final encrypted letter (Rodriguez-Clark, 2013).

So for the first letter of my text to be encrypted, 'G', I look to row B. 'G' encrypts to column F on row B, so letter 'G' is encrypted to 'H'.

Following this procedure, the encrypted message looks like this:

HMJGI SVSVR JKYAE DPSRV MQETI GIYEI OQTUL BJVQI OUREO UJVFY DPKHR
EQTUL BJVNY DPMVI MMJRL FVJJY SLZTO FQKRR HMJRL FVVFL BBMVI MAGNW
TOVZE DPKHR EQTUL PNWRH BAJVG IJRYH XQVQI SHLEY DSBRL SMEXE OV

The Vigenère square and modular arithmetic

An algebraic understanding of the Vignère Square uses modular arithmetic, because the letters of the key word 'wrap around' and start again when the word has ended (Sutherland, 2005).

A good example of modular arithmetic would be a clock face. The numbers go from 1 to 12, but '13 o'clock' effectively becomes 1 again (Neale, 2005). Similarly, after another loop of 12, '25 o'clock' is at the same position as 1 o'clock on the clock face (Neale, 2005).

This means that the remainder when you divide 13 or 25 by 12 is 1.
Mathematically, this is written as 13 = 1 mod 12, or 25 = 1 mod 12 (Neale, 2005).

When looking at the Vigenère Square, it is useful to work in, not mod 12, but mod 26, because there are 26 letters in the alphabet.

If we take the letters A – Z of the alphabet as corresponding to the numbers 0 – 25, then we can create a table like this:

A	B	C	D	E	F	G	H	I	J	K	L	M	N	O	P	Q	R	S	T	U	V	W	X	Y	Z
0	1	2	3	4	5	6	7	8	9	10	11	12	13	14	15	16	17	18	19	20	21	22	23	24	25

Using modular arithmetic and the numbers corresponding to the letters of the ciphertext, we can then calculate the number that corresponds to each letter of the original message using the following formula:

M = (C – K) mod 26 (Wikipedia, 2016)

M = number corresponding to letter of original message
C = number corresponding to letter of cipher text
K = number corresponding to letter of keyword

Letter of original message = (H – B) mod 26
(7 – 1) mod 26 = 6 mod 26 = 6 = G

Letter of original message = (M – I) mod26
(12 – 8) mod 26 = 4 mod 26 = 4 = E

Modular arithmetic can therefore be used to decrypt the Vigenère cipher if the key is known.

The Kasiski Examination

Using frequency analysis to decrypt the Vigenère cipher is ineffective because the most frequent letter will be encrypted as different letters throughout the message. The main weakness of the Vigenère cipher, however, is its key word.

The Kasiski Test takes advantage of the fact that repeated words in a message may by chance be encrypted with the same key letters (Rodriguez-Clark, 2013).

In the example below, I have selected the longest repeated words because they are less likely to be coincidental repetitions.

HMJGISVSVRJKYAEDPSRVMQETIGIYEIOQTULBJVQIOUREOUJVFYDPKHREQTULBJ
VNYDPMVIMMJRLFVJJYSLZTOFQKRRHMJRLFVVFLBBMVIMAGNWTOVZEDPKHREQ
TULPNWRHBAJVGIJRYHXQVQISHLEYDSBRLSMEXEOV

All the factors of the distance between the two repeated words are possible lengths of the key word (Rodriguez-Clark, 2013).

The distance between the first appearance of DPKHREQTUL and its repetition is 65. This suggests that the key word is 1, 5, 13 or 65 letters long. The most plausible option seems to be 5, as a key word of 1 or 3 letters is very short, and a key word of 65 letters is impossibly long.

The distance between the first appearance of MJRLFV and its repetition is 20.
The distance between the first appearance of LBJV and its repetition is 25.

5 is also a factor of both 20 and 25; therefore, the length of the key word is 5. If the length of the key is known, then the encrypted message can be treated as a number of interwoven Caesar Ciphers that are relatively easy to crack (Cryptomuseum.com, 2015).

The Friedman Test

In 1922, William Friedman published a statistical test that can determine whether a cipher is mono- or polyalphabetic, and for polyalphabetic ciphers such as the Vigenère cipher, can estimate the length of the key word (Christensen, 2015). This method is called the Friedman test and calculates the index of coincidence of a cipher text. This is the probability of randomly choosing two letters that are the same from the ciphertext (Cs.uri.edu, n.d.). Friedman noticed that when you draw two letters at random from a ciphertext, the index of coincidence is higher if the letters are drawn from the same alphabet than if they are drawn from different alphabets (Christensen, 2015).

These tables show the decimal frequency of monograms, or single letters, in German. I have combined the frequencies of 'a' and 'ä', of 'o' and 'ö', of 'u' and 'ü' and of 's' and 'ß', as a 26-letter alphabet doesn't cater for these additional characters.

Letter	Frequency
A	0.0688
B	0.0221
C	0.0271
D	0.0492
E	0.1588
F	0.0180
G	0.0302
H	0.0411
I	0.0760

Letter	Frequency
J	0.0027
K	0.0150
L	0.0372
M	0.0275
N	0.0959
O	0.0299
P	0.0106
Q	0.0004
R	0.0771

Letter	Frequency
S	0.0656
T	0.0643
U	0.0439
V	0.0094
W	0.0140
X	0.0007
Y	0.0013
Z	0.0122

(Lyons, 2012)

To find the probability of picking two of a particular letter, the probability of choosing one letter is squared. For a text in plaintext German, the probability of picking two letters that are the same, or the index of coincidence, is therefore:

$(0.0688)^2 + (0.0221)^2 + (0.0271)^2 + ... + (0.0122)^2 = 0.07161616 \approx 0.0716$

If the ciphertext were produced using a monoalphabetic cipher, then the index of coincidence would be near to 0.0716, because this type of cipher uses only a single alphabet, so the frequencies of letters in the ciphertext will be about the same as for plaintext German, just in a different order (Christensen, 2015).

However, if more than one alphabet is used, such as in a polyalphabetic cipher like the Vigenère cipher, then the frequencies of the letters should be more nearly uniform (Christensen, 2015), because the most frequent letters in plaintext German will be distributed among a number of different encoded letters.
The probability of choosing any given letter from the German alphabet (if we combine special characters with normal letters to create a 26-letter alphabet) is $\frac{1}{26}$, so the probability of choosing a pair of the same letters is $(\frac{1}{26})^2$.

Index of coincidence = $26 (\frac{1}{26})^2 = \frac{1}{26} = 0.03846153846 \approx 0.0385$ (Christensen, 2015)

Therefore, if the index of coincidence is closer to 0.0716, then the cipher is more likely to be monoalphabetic, but if it is closer to 0.0385, it is more likely to be polyalphabetic.

Friedman's formula to calculate the index of coincidence is:

$I = \frac{\sum_{i=0}^{25} n_i(n_i-1)}{N(N-1)}$ (www.cs.uri.edu, u.d.), where:
N = total number of letters in ciphertext
n_i = number of times letter i appears in the text (where i = A, B C...Z)

The frequencies of letters in the ciphertext on page 8 are communicated in the table below. There are 162 letters in the text in total.

A	B	C	D	E	F	G	H	I	J	K	L	M	N	O	P	Q	R	S	T	U	V	W	X	Y	Z
3	6	0	5	10	5	4	7	9	11	4	9	9	3	6	5	8	12	7	6	5	15	2	2	7	2

$\frac{\sum_{i=0}^{25} n_i(n_i-1)}{162(162-1)} = \frac{3(3-1)+6(12-1)+0(0-1)...+2(2-1)}{162(162-1)} = 0.0437466452 \approx 0.0437$

The index of coincidence for this text is approximately 0.0437, indicating that it is probably a polyalphabetic cipher, because 0.0437 is closer to 0.0385 than to 0.0716.

The table below is often used to relate the index of coincidence to possible lengths of the key word of a polyalphabetic cipher. It suggests that the key word is 5 letters long.

Estimated length of keyword	Index of coincidence
1	0.0660
2	0.0520
3	0.0473
4	0.0449
5	0.0435
6	0.0426
7	0.0419
8	0.0414
9	0.0410
10	0.0407
∞	0.0388

Figure 9

Once the index of coincidence has been calculated, an additional formula can be used to calculate the approximate length of the key word.

$$h \approx \frac{(E_s - \frac{1}{n})k}{(k-1)\Phi T - k\left(\frac{1}{n}\right) + E_s} \text{ (Ebermann, 2011)}$$

E_s = expected index of coincidence for German
ΦT = total index of coincidence of ciphertext
n = alphabet size
k = text length

$$h \approx \frac{(0.0716 - \frac{1}{26})162}{((162-1)0.0437) - \left(162\left(\frac{1}{26}\right)\right) + 0.0716} = 6.075456521 \approx 6.075$$

This formula suggests that the key word is 5 or 6 letters long.

Bayes' Theorem and Enigma

I have discovered that it would have been impossible for cryptanalysts to test all of the possible combinations on the Enigma machine (Population Genetics Graduate School, n.d.). However, Bayesian probabilities were used in Bletchley Park to reduce the number of possible rotor settings.

The principle behind Banburismus is relatively similar to Friedman's index of coincidence (Wikipedia, 2015). If two sentences (in English or German) are lined up one above the other, and a count is made of how often a letter in the first text is the same as the letter in the text below it, there will be more matches than if the sentences were random collections of letters (Wikipedia, 2015).

I used an online Enigma code simulator to encrypt two messages (Appendix D, E) with the same settings. The respective starting positions of the three rotors were BUL.

I compared these two messages for letter yields, highlighting matching characters in red.

LSCWINGYSONWLKPUPHKDQYICCXOOEPBATIRUFHEJLPZIUNMQPGJZQEEBERBMQNMAUUXYEFQBGCWWHMSZQ
LUAMIWICGCTNPSTAMBNQABJWOCJUHKTZJKGUYZICEJEEQUHDXBGXONEVNOHQZOAWQNAESWQUSGROBDZWQ

In 81 characters, there are 5 single-character repeats. This shows that two messages in the same language encrypted with the same Enigma settings have a higher repeat rate than the rate of 1 in 26 that we would expect for a comparison between random strings of letters.

I encrypted the second message again using BUV for the respective rotor starting positions. I compared this with the first message to count the number of repeats.

LSCWINGYSONWLKPUPHKDQYICCXOOEPBATIRUFHEJLPZIUNMQPGJZQEEBERBMQNMAUUXYEFQBGCWWHMSZQF
HNKRTCMDYQXQLWCHFYDVTETHJRIETVNWNGQXXIFCVLSIBRNEUUJAXACYKXJYATNXMOSNLGXNZIIDZEHVEF

In 82 characters, there are 3 monogram repeats. Because the two messages are not enciphered at the same settings of the Enigma machine, comparing them for yields gives us a value closer to that for texts made up of letters in random order (Hosgood, 2002), as $\frac{3}{82}$ is close to $\frac{1}{26}$.

The cryptanalysts at Bletchley would compare messages like those above, where the first two letters of the indicator are the same, at all possible offsets to look for promising repeat rates, as this could provide information as to how far apart the rotor settings of the Enigma machine were when the messages were enciphered (Hosgood, 2002).

The two messages can be aligned and one can be moved up to 25 places to the left or right, because there are 26 possible rotor starting positions. Therefore, there are 50 possible alignments that need to be tested (Simpson, 2010) One of these 50 overlaps must correspond to the messages having, not just the first two rotor settings in common, but also the third (Simpson, 2010).

Turing used the idea of the index of coincidence to suggest that the overlap with the highest number of repeats could indicate where letters in the original plaintexts were the same (Simpson, 2010).

I have already calculated that the index of coincidence for standard German is approximately 0.0716, and that two letters picked independently at random have a repeat rate of 0.0385. Therefore, if the hypothesis of two messages being in 'true depth' (all the pairs in the overlap having been enciphered alike at that position (Simpson, 2010)) is true, the event of a matching pair occurring across the two messages will have a probability of 0.0716, and if the hypothesis is false, and the messages are not in true depth, then there will be a probability of 0.0385 of finding a matching pair.

Although the repeat rate for German naval messages was actually shown to be closer to $\frac{1}{17}$ (Alexander, c.1945), or 0.0588, than 0.0716, I wanted to use my calculated value to link the different parts of my exploration together.

Bayes' Theorem suggests that the likelihood of the hypothesis (that the messages are in 'true depth') occurring before the event, multiplied by a factor of $\frac{0.0716}{0.0385} = 1.86$, will give the likelihood of the hypothesis occurring after the event (Simpson, 2010).

The formula for Bayes' theorem is: P(A/B) = $\frac{P(A)P(B/A)}{P(B)}$ (Fannon, 2012)

Let P(A) be the probability that a matching pair occurs in 'true depth' text.
Let P(B) be the probability that a matching pair occurs in the ciphertext.

We wish to calculate the probability of A occurring given that B occurs, or P(A/B). The weight of the evidence provided by B in favour of A being true is given by the Bayes factor (Simpson, 2010). This is the probability of a matching pair occurring in 'true depth' messages, divided by the probability of a matching pair occurring in random messages.

$1.86^n \times P(B/A) = P(A/B)$

The event of a pair of letters between the two messages not matching will therefore have a probability of (1 - 0.0716) if the hypothesis is true and (1 - 0.0385), or $\frac{25}{26}$, if the hypothesis is false (Simpson, 2010). This gives a factor of: $\frac{1-0.0716}{1-0.0385} = 0.967$.

The events of a pair of letters either matching or not matching are independent, so multiplying their factors together results in a factor for the whole alignment (Simpson, 2010).

LSCWINGYSONWLKPUPHKDQYICCXOOEPBATIRUFHEJLPZIUNMQPGJZQEEBERBMQNMAUUXYEFQBGCWWHMSZQF
HNKRTCMDYQXQLWCHFYDVTETHJRIETVNWNGQXXIFCVLSIBRNEUUJAXACYKXJYATNXMOSNLGXNZIIDZEHVEF

To use the example above, an overlap of 72 letters between these two messages resulted in 4 matching pairs and 68 non-matching pairs. These numbers are used as indices for the factors to calculate an overall factor for P(A/B) (Simpson, 2010).

$1.86^4 \times 0.967^{68} = 1.22194600124 \approx 1.22$

The prior odds of the hypothesis occurring were 1:49, as there were 50 equally likely possible alignments (Simpson, 2010). I have therefore divided the factor by 49 to calculate the posterior odds on A being true.

$\frac{49}{1.22} = 40.0999716437 \approx 40$

The posterior odds are 1:40. This suggests that, although the odds on the hypothesis have been slightly increased, the two messages are not very likely to be in 'true depth' when aligned at this particular placement.

At Bletchley Park, the overlaps between messages were recorded on paper sheets printed with vertical A-Z alphabets. (Simpson, 2010) For each letter in the encrypted message, a hole was punched in the columns in order (Wikipedia, 2015). The same was done with the second message, and then the two sheets were aligned above a lit background (Simpson, 2010). The number of times that the punched holes were visibly matching would be recorded. The sheets were called 'banburies', as they were printed in Banbury, and thus the whole process was named Banburismus (Wikipedia, 2015).

Conclusion

In my investigation, I researched the nature of several different codes and attempted to demonstrate their solving. This gave me greater insight into the concepts behind the Enigma machine and Banburismus, the method of decoding Enigma that was used by the cryptanalysts at Bletchley Park. I calculated how many possible Enigma settings there were, and then proceeded to investigate the mathematics of different codes and their solutions. Looking at frequency distributions of letters and the index of coincidence led me onto Bayes' theorem, which allowed me to link back to the start of my exploration in researching the mathematical principles behind Banburismus and how the odds on a hypothesis being correct and two messages being in 'true depth' are calculated.

I worked within the mathematical areas of probability, frequency distributions and modular arithmetic. I applied probability to Enigma by calculating the number of settings using the binomial coefficient, and also by working with the formulas for index of coincidence and Bayes' Theorem. I conducted my own frequency analysis and used the results to solve both simple ciphers and as a basis for the index of coincidence formula. Equally, the application of modular arithmetic to the Caesar and Vigenère ciphers was a useful tool to understand the looping nature of the alphabets in polyalphabetic ciphers.

Evaluation

This investigation was a challenge for me, as discovering that the subject I initially wanted to research was extremely complicated, I decided to take a slightly different path. This ended up being more useful in that I gained a more rounded understanding of ciphers as a whole, which is important because, in the history of coding, each cipher tends to be a slightly more complex development of the one that came before it, as each creator attempts to introduce an unbreakable cipher. The Enigma code was based on the ciphers than came before (from Caesar to Vigenère), and investigation into all of these ciphers helped me to understand that which was the most complex code yet at its time of usage.

The investigation was helpful to me in that breaking down a complex idea into a number of simpler codes helped me to better understand Banburismus when I returned to it at the end of the project. I don't think that I would have been able to understand Bayes' theorem if it hadn't been for my prior research which put it in context. For example, researching simpler polyalphabetic ciphers like the Vigenère cipher allowed me to understand the concept of the index of coincidence, which is also fundamental to the understanding of Banburismus. Looking at ciphers as rotating wheels also helped me to comprehend the nature of the rotors on the Enigma machine, and how their rotation makes for an almost infinite number of possible settings.

I also received much insight into the relevance of maths in cipher cracking. I learned how an important part of decrypting a cipher is the use of probability to reduce the number of possible solutions. I had not previously realised the fundamentality of probability to the area of ciphers. I also discovered how simple frequency analysis of letters can be input into more complex formulas to establish concrete facts about the nature of a code. I found it interesting to see the real-life application of probability in this way, and also enjoyed learning about modular arithmetic, which is not covered in the IB Standard Level course. Investigating the Kasiski Examination and the Friedman test was particularly interesting because I discovered formulas that fitted in with what I wanted to investigate.

A limitation of my investigation is that, due to my initial restricted understanding of ciphers, I had to start by researching the Caesar cipher before I could move on to more complex codes. This meant that a good deal of the investigation was taken up by maths of a simple level. In order to improve this, I would change the investigation by spending less time on the Caesar cipher and devoting more time to understanding the process of Banburismus, which I didn't research in as much depth as I would have liked.

Another limitation of the investigation is that some of the German text that I encrypted was not very realistic in the context of a naval World War Two setting. If I had more time, I would investigate the terminology and sentence structure that was commonly used in messages sent by code, and incorporate this into my code simulations to make them more realistic and closer to Enigma. I did try to do this towards the end of the investigation, but would have liked to implement it throughout.

If I could continue with the investigation, I would investigate codes that followed Enigma, such as the one-time pad, which is theoretically impossible to crack if used correctly. It would be interesting to see how attempts at analysis have been thwarted and whether there is ever likely to be a method of cracking the one-time pad if people like Alan Turing and the cryptanalysts at Bletchley Park investigated it.

7. The 'Perfect' Dive for a Swimmer

(assessment extracted from pdf)

Introduction

Background Context

Swimming has always been my number one sport that I engaged in competitively since the age of 7. I have participated in several swim competitions, both local and international. However, in order to be selected for the international galas I had to train endlessly, increasing speed, stamina, strength and making sure my technique was perfect. I would only qualify to participate in the top international competitions in case I achieved the fastest personal best time; I was thus always interested in ways to improve my technique and speed.

In competitive swimming, swimmers must respect and follow the approved rules give by the Fédération Internationale de Natation (FINA, English: International Swimming Federation) to avoid disqualification. However, the one area in the actual race that can be manipulated and gives the opportunity to achieve a good starting point within the approved rules, from when the referee blows the whistle to when you complete the race, is the starting dive. A good dive off a starting block is a crucial step in gaining a comparative advantage from the start. The dive can give the swimmer the acceleration needed to get ahead of his or her competitors. In training, I was exposed to multiple different dives that included a long dive, diving further away from the wall, and a high dive, jumping high and diving close to the wall. In multiple time trials we were told as swimmers to experiment to see which type of dive best fit us and to stick to the dive with thought was faster. My personal average time measured over a 10 metres distance was $11s$. Nonetheless, I was never able to figure out what my dive style was and it caused me to dive very differently at all times. I was not able to describe the exact relationship between the type of dive, its height and angle, and the time it takes the swimmer to cross a distance in the shortest time.

Topic of Assessment

The topic of this assessment is to figure out what the optimal angle of a dive that will result in a swimmers personal best time in a 10 meter swim. Specifically, studying the projectile motion of a dive and using calculus to determine the angle that will provide a minimal time. As seen in Figure 1, diving at different angles would alter the time spent in

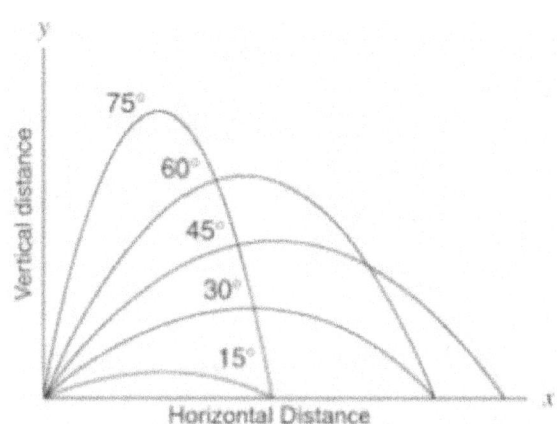

Fig.1. Brett A, *Projectile Motion and Horizontal Distance*, n.d.

the air and in the water. If the swimmer is faster in the water, than in the air, one would want to minimize the time spent during the dive, which will depend on the angle of dive. Therefore, I will be applying the knowledge I have learnt in class on differential calculus and kinematics and I will be learning new knowledge on projectile motion to determine the fastest optimal angle for the fastest dive.

Outline 'Projectile Motion'

The projectile motion of a dive is the motion through which a person dives towards the water's surface, while moving along a curved path under the action of gravity. However, the mathematics in projectile 'motion of a dive does not take into account the swimmers streamline posture and effects with air resistance.

Figure 2 represents the projectile motion of a dive. The velocity of the dive (v_D) has its horizontal (v_x) and vertical (v_y) components. The x axis represents the distance from the starting point and the

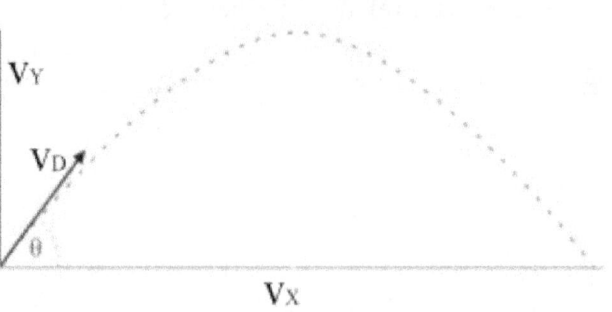

Fig. 2. Velocity in *Projectile Motion*, 2018

y axis represents the height of the jump of the swimmer. The parabola displayed in the graph represents the motion of the dive which concludes when the swimmer hits the water, which is where $y = 0$ and $x > 0$.

In the simplified diagram seen on Figure 3, the horizontal (v_x) and vertical (v_y) components of the velocity can be expressed as trigonometric equations. The angle of the dive (θ) represents the angle at which the swimmer jumps off the starting block. I am not indicating a specific angle as I am trying to determine the optimal

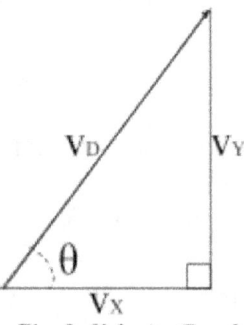

Fig. 3. *Velocity Graph*, 2018.

angle for a swimmer to dive at, crossing the distance of the dive (d_D), in relation to the velocity of the jump, to determine what angle can give the swimmer a faster motion.

The relationship between the sides of the triangle and the angle θ can be expressed as $sin\theta = \frac{Opposite}{Hypotenuse} = \frac{v_y}{v_D}$ and $cos\theta = \frac{Adjacent}{Hypotenuse} = \frac{v_x}{v_D}$ which can be rearranged to: $v_x = v_D cos\theta$ and $v_y = v_D sin\theta$.

Determine Distance of Dive (d_D)

After figuring out the individual components of the velocity (vertical and horizontal) the next step is determining the distance of the dive (d_D), which involves the velocity formula ($velocity = \frac{distance}{time}$) rearranged for distance: $distance = velocity \times time$.

In order to determine the equation of the distance of the dive (d_D), we look at both the horizontal distance (d_D) and use the concluding height of the dive (h_D) to determine the time it takes from the moment the swimmer jumps to the moment they are fully submerged below the water's surface (t_D). This is done by substituting the horizontal (v_x) and vertical (v_y) components of the velocity into the formula for finding distance.

The equation for the horizontal distance of the dive (d_D) is derived by substituting the horizontal (v_x) component of velocity into the distance formula: $d_D = v_x \times t_D$. As previously outlined, $v_x = v_D \cos\theta$ which means $d_D = v_D t_D \cos\theta$.

Similarly, to derive the vertical distance of the dive (h_D), I substituted in the vertical (v_y) component of velocity into the equation: $h_D = v_y \times t_D$. As previously outlined, $v_y = v_D \sin\theta$ which means $h_D = v_D t_D \sin\theta$.

Influence of Gravity on h_D

The equation for the vertical distance (h_D) requires to take into account the effects of gravity ($g = 9.8\ m/s/s$). This is because the swimmer is pulled downwards towards the surface of the water by the gravitational force after diving off the starting block. Without taking gravity into consideration, the obtained data would be distorted as the force is what pulls us back to the Earth's surface and without it we would be floating in space. Gravity can be incorporated by adding the term $-\frac{1}{2}gt_D^2$ (Kurtus) to the equation of the vertical displacement. Therefore the final equation for the vertical distance is $h_D = v_D t_D \sin\theta - \frac{1}{2}gt^2$.

Determine Time of Dive t_D

I will further refer to the two equations as:

Equation 1: $d_D = v_D t_D \cos\theta$ and

Equation 2: $h_D = v_D t_D \sin\theta - \frac{1}{2}gt^2$

In order to find the value of the horizontal distance (d_D), I need to determine the time of dive (t_D) using Equation 2.

For simplicity, it is assumed that the swimmer starts at an initial height $(h_i) = 0$, meaning that the swimmer is not jumping from an elevated height, and an initial time $(t_i) = 0$ thus $(0, 0)$. In my context, this would be jumping of the edge of the pool rather than off a diving block.

It is also assumed that h_D, the height of the swimmer at which he/she hits the water, is equal to the initial height h_i and is equal to 0. t_D is the time of dive, at which the swimmer plunges in the water, that can be derived from the Equation 2 assuming $h_D = 0$.

The obtained t_D will then be substituted into Equation 1, to find the total distance covered in the air (d_D).

The swimmer's horizontal distance moves forward on the x axis and the vertical distance takes form of a parabola starting from h_i and ending the dive when the swimmer hits the water $h_D = 0$.

$$\Rightarrow h_D = 0 = v_D t_D \sin\theta - \frac{1}{2}gt^2$$

Add both sides by $\frac{1}{2}gt^2$ in order to remove all negatives from the equation.

$$\Rightarrow \frac{1}{2}gt_D^2 = v_D t_D \sin\theta$$

Divide both sides by t as it is a common factor.

$$\Rightarrow \tfrac{1}{2} g t_D = v_D \sin\theta$$

Divide both sides by $\tfrac{1}{2}g$ in order to seclude t.

$$\Rightarrow t_D = \frac{v_D \sin\theta}{\tfrac{1}{2}g}$$

$$\Rightarrow t_D = \frac{2 v_D \sin\theta}{g}$$

$t_D = \frac{2 v_D \sin\theta}{g}$ is the expression of the time it takes the swimmer to cross the distance from the jump (0,0) to the moment he or she hits the water, as a function of the angle θ the swimmer takes off at and the initial velocity of the dive (v_D).

To find the horizontal distance of the dive (d_D), I am going to substitute the equation for t_D into Equation 1:

$$\Rightarrow d_D = v_D t_D \cos\theta$$

$$\Rightarrow d_D = v_D \cos\theta \left(\frac{2 v_D \sin\theta}{g} \right)$$

Expand and simplify:

$$\Rightarrow d_D = \frac{v_D \cos\theta \times 2 v_D \sin\theta}{g}$$

$$\Rightarrow d_D = \frac{v_D^2 \, 2 \sin\theta \cos\theta}{g}$$

Simplify using the double angle formula:

$$\Rightarrow d_D = \frac{v_D^2 \sin(2\theta)}{g}.$$

Determine Total Time $T(\theta)$

The total time, $T(\theta)$, of the 10 meter distance is found through the sum of the time of dive (t_D) and the time of swim (t_S), $T(\theta) = t_D + t_S$.

Express t_D and t_S

t_D has previously been expressed as $t_D = \frac{2v_D \sin\theta}{g}$.

In order to find t_S, I will assume that the swimmer dove into the water and immediately started with a constant velocity of swim (v_S). This assumption was made in order to simplify the equation and focus specifically on the dive. Therefore, to solve for t_S I will use the velocity formula, but rearranged for time:

$$\Rightarrow time = \frac{distance}{velocity}$$

In order to solve for time of swim (t_S), we use the velocity of swim (v_S) and introduce the distance of swim (d_S). t_S can therefore be expressed as $t_S = \frac{d_S}{v_S}$.

Determine d_S

In order to determine d_S we look at the total distance ($D(\theta)$). $D(\theta)$ is expressed as the sum of the distance of swim (d_s) and the distance of dive (d_D):

$$\Rightarrow D(\theta) = d_D + d_S.$$

As I said earlier, we will be looking at the time of a 10 meter distance, therefore:

$$\Rightarrow D(\theta) = 10$$

$$\Rightarrow d_S = 10 - d_D$$

I can now substitute in the values for t_D and t_S to determine the function for $T(\theta)$.

$$\Rightarrow T(\theta) = t_D + t_S.$$

$$\Rightarrow T(\theta) = \frac{2v_D \sin\theta}{g} + \frac{10 - d_D}{v_S}$$

d_D has previously been expressed as $d_D = \frac{v_D^2 \sin(2\theta)}{g}$, substitute it in;

$$\Rightarrow T(\theta) = \frac{2v_D \sin\theta}{g} + \frac{\left(10 - \frac{v_D^2 \sin(2\theta)}{g}\right)}{v_S}$$

This function can be used to measure the total time $T(\theta)$ when the sum of the distance of dive (d_D) and distance of swim (d_S) totals a distance of 10m as a function of the angle.

Establish a Formula for the Optimal Angle θ

I want to find the maximum value of the function for the total time $(T(\theta))$ and in order to do this, the next step would be to find the derivative of $T(\theta)$ by finding the values when the derivative equals zero.

$$T(\theta) = \frac{2v_D \sin\theta}{g} + \frac{\left(10 - \frac{v_D^2 \sin(2\theta)}{g}\right)}{v_S}$$

Derivative of the function for total time $T(\theta)$:

$$\Rightarrow \frac{dT}{d\theta} = \frac{2v_D}{g}\cos\theta - \frac{2v_D^2 \cos(2\theta)}{v_S g} = 0$$

Divide both sides by $\frac{2v}{g}$ because it is a common factor:

$$\Rightarrow 0 = \cos\theta - \frac{v_D \cos(2\theta)}{v_S}$$

Implement the double angle formulae: $\cos(2\theta) = 2\cos^2(\theta) - 1$

$$\Rightarrow 0 = \cos\theta - \frac{v_D}{v_S}(2\cos^2\theta - 1)$$

$$\Rightarrow 0 = \cos\theta - \frac{2v_D}{v_S}\cos^2\theta + \frac{v_D}{v_S}$$

In order to simplify the next few calculations I am substituting $\cos\theta$ with x.

$$\Rightarrow 0 = x - \frac{2v_D}{v_S}x^2 + \frac{v_D}{v_S}$$

The above equation can be reorganized into the standard form of a quadratic equation.

$$\Rightarrow ax^2 + bx + c$$

$$\rightarrow a = -\frac{2v_D}{v_S}, \, b = 1 \text{ and } c = \frac{v_D}{v_S}$$

$$\Rightarrow 0 = -\frac{2v_D}{v_S}x^2 + x + \frac{v_D}{v_S}$$

Use the quadratic formula to solve for x.

$$\Rightarrow x = \frac{-b \pm \sqrt{b^2 - 4ac}}{2a}$$

$$\Rightarrow x = \frac{-1 \pm \sqrt{1^2 - 4\left(-2\frac{v_D}{v_S}\right)\left(\frac{v_D}{v_S}\right)}}{2\left(-2\frac{v_D}{v_S}\right)}$$

Expand the brackets:

$$\Rightarrow x = \frac{-1 \pm \sqrt{1 + 8\frac{v_D^2}{v_S^2}}}{-4\frac{v_D}{v_S}}$$

$$\Rightarrow x = \cos\theta = \frac{-1 \pm \sqrt{1 + 8\frac{v_D^2}{v_S^2}}}{-4\frac{v_D}{v_S}}$$

Remove the double fraction by multiplying the equation by the reciprocal of the denominator

$$\Rightarrow \cos\theta = -\frac{v_S}{4v_D}\left(-1 \pm \sqrt{1 + 8\frac{v_D^2}{v_S^2}}\right)$$

Cancel out the negative signs, two negatives make a positive.

$$\Rightarrow \cos\theta = \frac{v_S}{4v_D}\left(1 \pm \sqrt{1 + 8\frac{v_D^2}{v_S^2}}\right)$$

Rearrange formula to solve for θ, input the inverse cosine function:

$$\Rightarrow \theta = \cos^{-1}\left[\frac{v_S}{4v_D}\left(1 \pm \sqrt{1 + 8\frac{v_D^2}{v_S^2}}\right)\right]$$

This equation will allow me to see what the angle θ, has to be so that it takes me the least amount of time to complete the 10m. Now I can input the values of the velocities in order to solve for the angle of the dive.

Testing in Real Life Context

In order to test the formula for θ, we will input random variables for v_D and v_S. To increase the reliability and real life context, I derived the velocity of swim using existing speeds of professional swimmers:

- the world record time on men's 50 meter freestyle long course, set by Brazilian swimmer Cesar Cielo in 2009, who swam with a time 20. 91 seconds for an average speed of the race 2.39 m/s (Allein) and
- the Olympic record time on women's 400 meter freestyle long course, set by US swimmer Katie Ledecky in 2016, who swam with a time of 236.46 seconds for an average speed of 1. 6916 m/s. (Lutz)

Knowing these are professional swimmers, I assume that an amature swimmer's time would be relatively slower. My personal average time measured when I was experimenting with the different angles of jump and swimming techniques, was $v_S = 0.86\ ms^{-1}$ so for this test it is realistic to assume the value would be $v_S = 2ms^{-1}$. I chose this value because it is a rounded middle value between my measured value and the researched competitive swimmers' values; I also want to keep the numbers simple so that the mathematics will produce feasible and meaningful results, that will help me assess the accuracy of the formula.

Due to the unknown corresponding velocity of dive for the two competitive swimmers, I chose a random variable of $1ms^{-1}$. Nonetheless, I will test the formula twice. Test 2 will include a different velocity of dive, determined by the conclusion made in Test 1, to gain a high level of accuracy in the results.

The conclusion will be determined by checking if the result or results for $cos(\theta)$ are between -1 and 1, since those are the minimum and maximum values of cosine respectively. I will then use the inverse cosine rule to solve for θ using radians.

Test 1

- $v_D = 1ms^{-1}$
- $v_S = 2ms^{-1}$
- $\frac{v}{v_S} = \frac{2}{3}$

$$\Rightarrow \theta = \cos^{-1}\left[\frac{v_S}{4v_D}\left(1 \pm \sqrt{1 + 8\frac{v_D^2}{v_S^2}}\right)\right]$$

$$\Rightarrow \theta = \cos^{-1}\left[\frac{2}{4(1)}\left(1 \pm \sqrt{1 + 8\frac{1^2}{2^2}}\right)\right]$$

Simplify further to determine the results:

$$\Rightarrow \cos(\theta) = \left[\frac{2}{4(1)}\left(1 \pm \sqrt{1 + 8\frac{1^2}{2^2}}\right)\right]$$

$$\Rightarrow \frac{2}{4}\left(1 + \sqrt{1 + 8\frac{1}{4}}\right) \qquad \Rightarrow \frac{2}{4}\left(1 - \sqrt{1 + 8\frac{1}{4}}\right)$$

Results:

$$\Rightarrow 1.36603 \neq \cos(\theta) \qquad \Rightarrow -0.36603 = \cos(\theta)$$

$$\Rightarrow \cos^{-1}(-0.36603) \approx 1.94^c \approx 111.5°$$

These variables do not work as when I added I obtained a value greater than 1, which was outside the range of $\cos(\theta)$, and when subtracting I obtained an angle where the swimmer would be jumping backwards.

The problem was that $\frac{v_S}{4v_D}\left(1 + \sqrt{1 + 8\frac{v_D^2}{v_S^2}}\right)$ was greater than 1, so in order to reason out the solution to the equation, I decided find what happens when the equation is equal to 1, which means jumping horizontally at angle of zero, even though this would be physically impossible to dive and instead is similar to pushing yourself off of the wall. However, I think it will help to determine what sort of value would be inputted for v_D and v_S and therefore I will assume $cos(0) = 1$.

Instead of giving particular values for the velocities, I am now giving the value for the simplest angle (0) and I will solve for v_D and v_S. I am considering only the plus solution for simplicity but the results should also apply when subtracting:

$$\Rightarrow \frac{v_S}{4v_D}\left(1 + \sqrt{1 + 8\frac{v_D^2}{v_S^2}}\right) = 1$$

Divide both sides by $\frac{v_S}{4v}$ and open up the brackets:

$$\Rightarrow 1 + \sqrt{1 + 8\frac{v_D^2}{v_S^2}} = \frac{4v_D}{v_S}$$

Subtract both sides by 1 and remove the square root by squaring both sides:

$$\Rightarrow \sqrt{1 + 8\frac{v_D^2}{v_S^2}} = \frac{4v_D}{v_S} - 1$$

$$\Rightarrow \left(\sqrt{1 + 8\frac{v_D^2}{v_S^2}}\right)^2 = \left(\frac{4v_D}{v_S} - 1\right)^2$$

$$\Rightarrow 1 + 8\frac{v_D^2}{v_S^2} = \left(\frac{4v_D}{v_S} - 1\right)^2$$

Use $(a-b)^2 = a^2 - 2ab + b^2$ for $\left(\frac{4v_D}{v_S} - 1\right)^2$:

$$\Rightarrow 1 + \frac{8v_D^2}{v_S^2} = \frac{16v_D^2}{v_S^2} - \frac{8v_D}{v_S} + 1$$

Subtract both sides by 1, move all the variables to one side to equal 0 and simplify:

$$\Rightarrow 0 = \frac{16v_D^2}{v_S^2} - \frac{8v_D^2}{v_S^2} - \frac{8v_D}{v_S}$$

$$\Rightarrow 0 = \frac{8v_D^2}{v_S^2} - \frac{8v_D}{v_S}$$

Factor out the common variables:

$$\Rightarrow 0 = \frac{8v_D}{v_S}\left(\frac{v_D}{v_S} - 1\right)$$

Results:

$$\Rightarrow 0 = \frac{8v_D}{v_S} \text{ or } \frac{v_D}{v_S} = 1$$

In interpreting these results I found that $0 = \frac{8v_D}{v_S}$ would not work as it implies the speed of dive would equal 0. However, when $\cos\theta = 1$, $\frac{v_D}{v_S} = 1$. To conclude Test 1, I can summarize that when $\frac{v_D}{v_S} < 1$, meaning when $v_D < v_S$, the results falls outside the range of $\cos(\theta)$, which is considered to be a non-real calculation. The implication of this is that velocity of dive should not be slower than velocity of swim. In order to prove the accuracy of this statement, I am going to find what happens when $v_D > v_S$.

Test 2

- $v_D = 4ms^{-1}$
- $v_S = 2ms^{-1}$
- $\frac{v_D}{v_S} = \frac{4}{3}$

$$\Rightarrow \theta = \cos^{-1}\left[\frac{v_S}{4v_D}\left(1 \pm \sqrt{1 + 8\frac{v_D^2}{v_S^2}}\right)\right]$$

$$\Rightarrow \theta = \cos^{-1}\left[\frac{2}{4(4)}\left(1 \pm \sqrt{1 + 8\frac{4^2}{2^2}}\right)\right]$$

Simplify further to determine the results of $\cos(\theta)$:

$$\Rightarrow \cos(\theta) = \left[\frac{2}{4(4)}\left(1 \pm \sqrt{1 + 8\frac{4^2}{2^2}}\right)\right]$$

$$\Rightarrow \frac{2}{16}\left(1 + \sqrt{1 + 8\frac{16}{4}}\right) \qquad \Rightarrow \frac{2}{16}\left(1 - \sqrt{1 + 8\frac{16}{4}}\right)$$

$$\Rightarrow \approx 0.84307 \qquad \Rightarrow \approx -0.59307$$

Both values are between -1 and 1 therefore to solve for θ:

$$\Rightarrow \theta = cos^{-1}(0.84307)$$

$$\Rightarrow \approx 0.567829^c \approx 32.5°$$

$$\Rightarrow \theta = cos^{-1}(-0.59307)$$

$$\Rightarrow \approx 2.20566^c \approx 126°$$

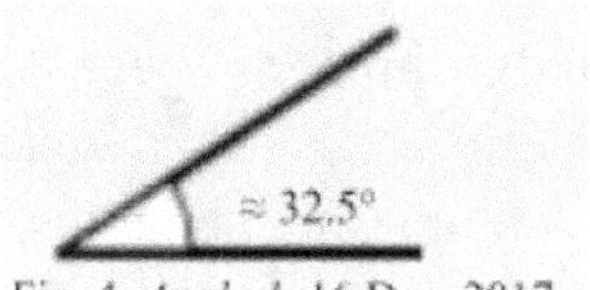

Fig. 4. *Angle 1*, 16 Dec. 2017.

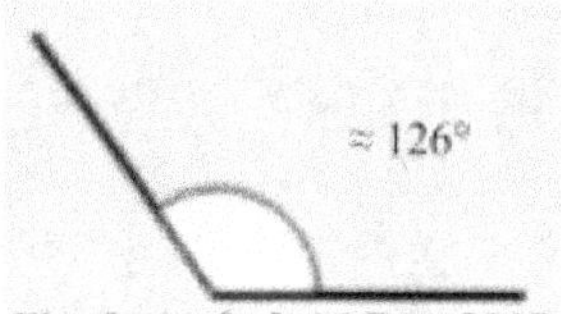

Fig. 5. *Angle 2*, 16 Dec. 2017.

After receiving the results of test 2, I can state that the addition option produces feasible results in determining the optimal angle. In reference to the context of the report focusing on a swimmers dive, 32.5° would be the optimal angle when the velocity of swim is $2ms^{-1}$ and the velocity of dive is $4ms^{-1}$. This because if we look at Figure 4 and Figure 5 the only one that follows an angle of a dive would be Figure 4, whereas Figure 5 would have the swimmer jumping backwards. The dive would thus be lower and further away from the wall as seen in Figure 6.

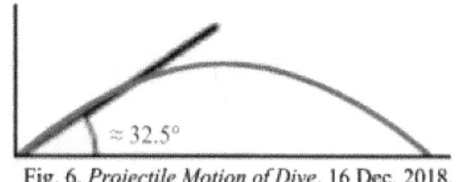

Fig. 6. *Projectile Motion of Dive*, 16 Dec. 2018.

These results lead me to believe that the formula for the optimal angle of a dive can be determined using the equation $\theta = cos^{-1}\left[\frac{v_S}{4v_D}\left(1 + \sqrt{1 + 8\frac{v_D^2}{v_S^2}}\right)\right]$.

Conclusion

When proving that v_D has to be greater than v_S mathematically, I realized that this is because if it wasn't, it would be better to just swim and omit the dive. In relation to my personal experience and my recorded times, I always achieved a faster time with the dive than with just pushing myself of the wall. When we jump, we do it to provide us with an extra push in the beginning so that before we start to swim we are already moving therefore, the dive has faster velocity as you are jumping off of the wall.

Through the mathematics of investigating projectile motion, differential calculus, kinematics and its relationship with dive of a swimmer, I believe that the equation that was formed is effective as it is dependent on the individual swimmer's speed of dive and swim. The formula represents a generalized relationship among the variables, I doesn't set a standard and can be used amongst every individual swimmer. The only condition is that the velocity of the dive is faster than the velocity of the swim. The formula to solve for the optimal angle, $\theta = cos^{-1}\left[\frac{v_S}{4v_D}\left(1+\sqrt{1+8\frac{v_D^2}{v_S^2}}\right)\right]$, provides the swimmer with the best angle they should try to dive at, and with that angle they can also calculate approximately how much time it would take to complete a 10 meter swim using the function $T(\theta) = \frac{2v_D \sin\theta}{g} + \frac{\left(10 - \frac{v_D^2 \sin(2\theta)}{g}\right)}{v_S}$.

I can assume that the total time $T(\theta)$ for any variables will reflect a goal between the time achieved by competitive swimmers and my own average time, and will be generally faster than we actually swim 10 meters in, nonetheless it provides any competitive swimmer with a goal and improvement in technique in order to reach that goal successfully. For example, using the data from Test 2, with θ being in radians:

$$\Rightarrow T(0.567829) = \frac{2(4)\sin(0.567829)}{9.8} + \frac{\left(10 - \frac{(4)^2 \sin(2 \times 0.567829)}{9.8}\right)}{3}$$

$$\Rightarrow T(0.567829) = 3.48$$

It should take approximately 3.48 seconds for the swimmer to reach 10 meters, which is a feasible time and good goal for time to gain a new personal best time.

However, a limitation of this investigation may be the omission of the external physical factors and other intervening variables in the actual dive done by a swimmer. These include the weight of a swimmer and the posture and body structure of the swimmer. A further limitation may have arisen due to assuming the water was the same level as the ground the swimmer was diving off; however, in competitive swimming, diving from a height is inevitable because of the use of diving blocks. These factors were omitted because they would increase the amount of added constants and affect the feasibility of the calculations. As a simplified model of a real life situation, I believe the report produced a useful conclusions to me, helping me to resolve the question that was puzzling to me every time I needed a good dive in a swim competition.

My conclusions can be further expanded by incorporating the elevated initial height. If you were to calculate the dive from a height you can add a α meters to equation 2 to get $y = v_D sin\theta t - \frac{1}{2}gt^2 + \alpha$ where alpha is the initial distance from which to jump and solve it following the same steps. Whereas the working out will be more complicated by adding this new variable, the results will most probably include a similar conclusion.

Works Cited

Allain, Rhett. "Olympic Physics: Swimming, Power and Setting Records." *Wired*, CNMN Collection, 4 Aug. 2012, www.wired.com/2012/08/olympics-physics-swimming/. Accessed 13 Dec. 2017.

Kurtus, Ron. "Gravity Displacement Equations for Falling Objects." *School of Champions*, 9 June 2011, www.school-for-champions.com/science/gravity_equations_falling_displacement.htm . Accessed 5 Aug. 2018.

Lutz, Rachel. "Katie Ledecky Wins First Individual Gold of 2016 Olympics in 400m Freestyle with World Record." *NBC Olympics*, NBC Universal, 8 Aug. 2016, www.nbcolympics.com/news/katie-ledecky-wins-first-individual-gold-2016-olympics-400m-freestyle-world-record. Accessed 13 Dec. 2017.

A, Brett. Projectile Motion and Horizontal Distance. Wikispace, Tangient, 2012, researchthetopic.wikispaces.com/What+is+projectile+motion%3F+-+R. Accessed 16 Dec. 2017.

www.ingramcontent.com/pod-product-compliance
Lightning Source LLC
Chambersburg PA
CBHW080026080526
44586CB00017B/2140